LAND AND THE CITY

997

LAND AND THE CITY

Patterns and processes of urban change

Philip Kivell

London and New York

For my family

First published 1993
by Routledge
11 New Fetter Lane, London EC4P 4EE

Simultaneously published in the USA and Canada
by Routledge
a division of Routledge, Chapman and Hall Inc.
29 West 35th Street, New York, NY 10001

© 1993 Philip Kivell

Typeset in [Scantext Baskerville] by
Leaper & Gard Limited, Bristol
Printed and bound in Great Britain by
Biddles Ltd, Guildford and King's Lynn

British Library Cataloguing in Publication Data

A catalogue reference for this book is available from the British Library

ISBN 0–415–08781–3 Hb
0–415–08782–1 Pb

Library of Congress Cataloging in Publication Data

Kivell, Philip.
    Land and the city : patterns and processes of urban change /
Philip Kivell.
        p.   cm. – (Geography and environment series)
    Includes bibliographical references (p.   ) and index.
    ISBN 0–415–08781–3 (HB). – ISBN 0–415–08782–1 (PB)
    1. Land use, Urban.   I. Title.   II. Series.
HD1391.K57   1993
333.77′13–dc20
                                        92–19286
                                        CIP

CONTENTS

CONTENTS

FIGURES

TABLES

PREFACE

This book is an attempt to provide a broadly based, yet succinct statement of land use patterns and processes in urban areas. It has been written in the conviction that land has been, and will continue to be, of the greatest importance in helping us to understand both the spatial patterning and the functioning of modern settlements and societies. Urban areas, although changing profoundly, remain of overwhelming and obvious importance in western societies.

Quantitatively, land given over to urban uses is of relatively minor importance, occupying typically no more than 5 to 10 per cent of the total in most of the developed nations. However, it has a significance far greater than this small share of total area implies, for in most of these countries between 75 and 90 per cent of the population live in urban locations. Most obviously land can be seen as a container of human communities and their economic activities, it provides the basic morphological elements, or the physical skeleton, of the city. In this it provides a strong reflection of the space needs of both past and present technologies and can be viewed as a record of evolving patterns of life. Urban land use is remarkably durable and in the central areas of many older cities morphological patterns which are many hundreds of years old may be retained. A second way in which urban land is important is as a source of power. Simple economic power may be gained from the ownership of valuable urban land, but a more subtle form of social power and status may also be exercised by individuals or groups who hold land in select locations. Third, land is the key to planning and control by government and other institutions. In this sense land use and ownership is inherently political. Finally, land is also intimately connected with environmental issues. Urban areas are vast consumers of resources and thus exert their influence over wide areas, but even viewed more narrowly it is clear that there is currently great concern over the quality of the urban environment in which such large numbers of people live.

Urban areas have been the subject of large amounts of academic analysis, especially recently, but the important land use aspects have invariably been neglected. One of the main purposes of this book is to cover those neglected areas.

Many processes affect the pattern and changing use of urban land, but two broad sets of processes and the interplay between them will be given particular attention. First, there are a number of market forces which include the competition between different urban activities, the changing locational needs of industry and commerce, the nature of the urban land value surface and changing transport technologies. These remain central to any understanding of urban land, and although they have been fairly fully documented elsewhere, the discussion of them, and their interplay with land issues has not been well integrated. Second, and of equal importance, is the operation of land use planning systems and the impact of government policies on the use and development of urban land.

Two main perspectives dominate the book, the first of which is geographical. Geographers have undertaken many detailed analyses of urban areas, but the focus has most commonly been upon economic or social structures and the underlying issues relating to land have often been omitted. Land has been, and still is, central to the discipline of geography, but many land use studies have examined land in general and have treated urban land as a residual use. Ironically, much of our understanding of urban land use patterns and structures is still informed by models, concepts and processes which are over half a century old. The second main perspective is that of town planning. Despite the fact that planning has been somewhat relegated by strong government beliefs in market forces in the past decade, and the fact that planning, as practised, concerns ostensibly economic or social issues, land use remains one of the keys to the whole process. Recent years have seen many new land policies introduced and, in many cases, it is land use policy which has been chosen as an important weapon with which to combat a variety of urban problems. Especially in the declining inner city areas, it is land development which has been used to spearhead urban regeneration.

The structure of the book is quite straightforward, and mostly it emphasises practical and applied land use issues in the developed nations of the world. The introductory chapter sets the basic urban context and discusses the importance of land in a number of different senses. The following chapter looks at the way in which land is allocated to different uses, particularly through market forces and public intervention, and it also critically examines a number of established models of urban land use. Chapter 3 examines the sources and nature of information about land use in urban areas and argues that our state of knowledge is both patchy and inadequate. Recent advances in geographical information systems have greatly enhanced our ability to analyse and interpret the statistics, but the raw data leaves much to be desired. In Chapter 4 some of this information is brought together in an attempt to summarise what we know about the extent, composition and current trends in urban land use. The complicated question of land ownership forms the focus of Chapter 5 and a distinction is made between private interests and the public sector. The behaviour of land owners is seen to be vital to the land development process and is important in shaping the city. The role of the public sector is further examined in Chapter 6 in the context of land policy. Some of the reasons for, and instruments of, land policy are discussed and a number of different national case studies are provided. In Chapter 7 the question of vacant and derelict land is explored, some of the reasons for its existence are listed and there is a brief discussion of land restoration policies.

Finally, Chapter 8 attempts to provide some conclusions, in particular by reviewing a number of important contemporary trends in urban development which have land use implications. It could, no doubt, be argued that any twenty-five-year period during this century has seen profound changes affecting big cities. The first quarter of the twentieth century witnessed massive changes in the social structure of British cities, political realignments in Europe and the emergence of the American city as a melting pot assimilating waves of migrants. The second quarter saw widespread economic recession, the emergence of many social problems and the inexorable beginnings of suburban growth. The third quarter was one of extensive economic growth and stability, the final maturing phase of urban manufacturing economies, the challenge of industrial growth in the cities of the Third World, the dismantling of Britain's overseas empire and the beginnings of the collapse of older, inner city communities. Against such a momentous background, it may be difficult to argue that the changes of the last quarter of the twentieth century are anything unusual. Yet it is clear, as Chapter 8 demonstrates, that profound changes are in train. The decline of manufacturing industry, continued suburbanisation and counterurbanisation, technological changes as we move into a post industrial era and new lifestyle preferences are not simply altering the city in detail. They are creating new, more dispersed forms of urban settlement in which our old concepts of the city have less and less relevance, and in which the urban fringe is becoming the new centre of activity. Above all this new wave of urban change is bringing with it far reaching implications for urban land use.

ACKNOWLEDGEMENTS

The author would like to acknowledge many kinds of help provided by a number of people in the preparation of this book. Several Keele colleagues provided moral and intellectual support. Particular thanks must go to Pauline Jones and May Bowers for handling the original typescript, to Andrew Lawrence for preparing the diagrams, to Michael Bradford for constructive comments on early drafts and to Michael Chisholm, whose collaboration on work leading to Chapter 7 was especially valued.

Chapter 3 originally appeared in Healey, M.J. (ed.) (1991) *Economic Activity and Land Use*, Harlow, Longman, and is reprinted here by kind permission of the publishers.

The author also gratefully acknowledges the following for permission to use or reproduce material:

Butterworth Heinemann (Table 4.11); Controller of HMSO (Figs 7.1, 7.2, Tables 2.1, 3.1, 4.10, 4.13, 7.1, 7.3, 7.4, 7.5); R. Davies (Fig. 2.5); *Financial Times* (Fig. 5.1); Y. Himiyama (Table 4.11); Hodder and Stoughton (Tables 4.7, 5.1), B.S. Hoyle (Fig. 7.5); Institute of British Geographers (Figs 5.2, 5.3, 5.4); London Docklands Development Corporation (Fig. 4.1); Macmillan Press (Fig. 2.3 part); Methuen and Co. (Fig. 2.3 part, Tables 4.6, 4.9); Ordnance Survey, Crown Copyright (Fig. 4.2), Northwestern University, Evanston, USA (Fig. 2.4); Resources for the Future, Washington D.C. Copyright (Table 4.8); Routledge (Figs 2.3 part, 2.6, 6.2, Tables 5.5, 6.1); J. Speake (Figs 4.4, 7.3, 7.4); Shepheard-Walwyn (Publishers) Ltd. (Table 4.6); The American Geographical Society, New York (Table 5.4); Urban Studies (Figs 2.7, 2.8).

Although every effort has been made to trace and contact copyright holders, this has not always been possible. We apologise for any apparent negligence.

Abbreviations

CORINE	Co-ordinated Information on the European Environment
DoE	Department of the Environment
GIS	Geographical Information Systems
ha	hectares
LAMIS	Local Authority Management Information System
NERC	National Environmental Research Council
NLUC	National Land Use Classification
OPCS	Office of Population Censuses and Surveys
SPOT	Satellite Probatoire d'Observation de la Terre/Le Systeme Pour l'Observation de la Terre

1

INTRODUCTION

In the organisation of its economies and the spatial patterning of its settlements, the developed world is overwhelmingly urban. Some of the oldest and
most extensively developed urban societies, notably Britain and the USA,
have never been ideologically enthusiastic about cities, but until the middle
of this century universal urban growth on a large scale was the norm. Today
over 70 per cent of the population of the developed world lives in urban areas
and an even higher proportion of economic, political and administrative
power is concentrated there. Even with recent population and employment
trends acting to disadvantage the largest cities in many developed countries,
the way of life and the organisation of economic activity is still dominantly
urban.

Within these urban settlements land is used and occupied in a remarkably concentrated manner. Different national definitions make exact
comparisons difficult, but some general patterns may easily be identified. In
the United States, for example, metropolitan areas house 75 per cent of the
population on just 1.5 per cent of the total land. In France fifty-eight *unites
urbaines*, each exceeding 100,000 people, between them accommodate 44
per cent of the nation's population on less than 1 per cent of the land. In
England and Wales, which are smaller and more densely urbanised than
most countries, 89 per cent of the population live in urban areas on 7.7 per
cent of the land (DoE 1988), and in Japan extreme urban concentration can
be observed in the three agglomerations of Tokyo, Osaka and Nagoya which
account for half of the total population on less than 2.5 per cent of the
national land surface. The United Nations Organisation's *Global Review of
Human Settlements* estimated in 1976 that Western Europe was the most
densely urbanised of the world's major regions with 3 per cent of its land
surface built upon. Other figures were 0.2 per cent for the USA/Canada and
Australia/New Zealand, and 0.4 per for Eastern Europe. Even if the world's
urban population lived at the low densities typical of North America, they
would cover less than 1 per cent of global land.

Such low figures as these give a misleading underestimate of the importance of urban land. Judged in terms of economic output or capital values,

1

urban land is of vastly greater significance. Even using the crude measure of the amount of land which is urbanised, it can be seen that in the economic core regions of the developed world the proportions are far higher. Within the south east of England for example, the urban coverage is approximately one-fifth of the total land.

In these highly urbanised regions, and elsewhere, urban land continues to be in great demand. This is despite, or perhaps because of, counterurbanisation trends whereby the largest cities are losing their domination over national economies and settlement systems. Many of the reasons for this will be taken up in greater detail in later chapters, but for the present a number of basic explanations for the growth in demand for land can be outlined.

Population growth in the developed nations is a prime cause of urban expansion. Put simply, more people consume more land, and for the most part this means more urban land. Crude population growth in many developed countries has been relatively modest in recent years, but it has played a part in the process of urban expansion. The population of the United Kingdom grew by only 10 per cent between 1951 and 1981, but this resulted in 5.5 million extra people. At typical suburban densities of 75 persons per hectare, this represented an additional 73,000 hectares of land for housing alone, equivalent to ten times the existing area of the city of Nottingham. Even in the absence of crude population growth, additional demands for urban housing have been generated by social and demographic changes such as divorce, changing marriage patterns and an ageing population. Each of these contributed to a reduction in average family size and a higher number of separate households.

Increasing personal affluence is a second major influence upon the demand for and disposition of urban land. Whatever the individual or local effects of the recent recession, there has been an overall upward trend in affluence. In western cities where the basic needs of life are already amply catered for, the effect of growing affluence has been to provide a further boost to the consumption of land. Mostly this involves rising living standards, translated into lower residential densities and an increasing number of motor vehicles, but it has also resulted in growing leisure time and a multiplication of leisure activities. Most of these activities demand land in, or close to the major urban population centres.

The uncertain financial climate of recent years has seen land and property re-emerge as the traditional safe hedge against inflation. Coupled with life-style preferences this has resulted in a rapid growth in home ownership, a trend which for financial as well as social reasons has been largely satisfied by extensive areas of low density individual family homes. The near certainty of medium to long term rises in the value of houses has been a particular attraction when the financial returns from other forms of investment have been unreliable.

The property boom has been facilitated by the availability of finance. This has come partly from growing personal affluence which has had its main impact in the residential sector. In Great Britain , for example, home ownership rose from 49 per cent of households in 1971, to 66 per cent in 1989. Tremendous growth has also taken place in property investment by institutions, such as pension funds, investment and insurance companies. These have been particularly attracted to land and property development in the commercial sector, notably through hotels, offices and the retail and leisure sectors, although 1992 saw a dramatic downturn in property.

Transport and communication changes represent another major explanatory variable in helping to account for the continuing demand for urban land. In particular the shift from rail to motor vehicle transport which occurred in the USA from the 1920s onwards, and in Europe rather later, enabled the much freer movement of people and goods from point to point rather than being confined to predetermined lines. This allowed a loosening of the urban fabric to take place and prompted new phases of urban development. At the same time, major new nodes emerged in the form of motorway interchanges and junctions, and international airports which presented new locational opportunities for economic activity, often on the outskirts of cities where land was relatively cheap. These opportunities were rapidly seized by developers of office and retail complexes, science parks and high technology industrial estates, all of which created new urban forms. More recently the plethora of telecommunication and other electronic advances has given a further twist to the locational patterns of economic activity, to the point where some of the traditional connotations of urban have all but lost their meaning.

Urban land use must thus be seen as a constantly evolving pattern rather than a static entity, even so the past quarter of a century has seen more rapid and profound change than at any other time in recent history. It is also important to view land as a multifaceted aspect of urban development, not simply serving as a neutral space or container of activities and objects but as an intrinsic part of virtually all aspects of urban life. Above all, land is the key to understanding two important aspects of urban development. First, it is vital in explaining the shape, layout and growth of urban forms. Second, it is at the centre of the city's activities, influencing economic development conferring power and determining the relationships between different social groups and activities.

LAND AS URBAN MORPHOLOGY

Individual cities display morphologies and land use patterns which range from the very formal and carefully ordered to apparently haphazard collections of buildings, spaces and activities. The precise pattern is determined by a multiplicity of factors including the age, style and scale of development, the needs for different kinds of land and the nature of its ownership. The economic explanation of the land-use pattern must incorporate forces which extend far beyond the city's local boundaries. As Carter suggested (1983: 114): 'the plan and built form of the town are direct reflections of the nature of culture on the large scale ... the town epitomises in its physical nature the complex of political, economic, and social forces which characterised the period of its creation'.

As examples Carter contrasted the towns of Renaissance and Baroque Europe with their formal layout, classical architecture and defensive needs with the very different but equally rigid morphologies of the North American gridiron plan. Both were shaped by strong, centralised planning influences; in the former case the aristocratic land owners and state military architects, in the latter case by the Land Ordinance established to regulate the settlement of North America after 1785. The rationale behind the European designs can be seen in terms of the aesthetic appeal of formality, the nature and scale of contemporary warfare and the desire for pomp and ceremony. The American example, on the other hand, has been explained (Stanislawski 1946), as a reflection of the democratic principles of the time and the desire to divide and allocate land on broadly egalitarian lines. This may be true, but undoubtedly the system also owed much to the needs of simplicity and expediency in laying out numerous new settlements on largely virgin land. As with most morphological designs, these have remained remarkably enduring features of the towns in which they were applied.

A more marked contrast may be found by examining the morphologies and forms of land development which typified the explosive growth of industrial urban settlements in the nineteenth century. These forms are particularly widespread in Britain and the rest of Europe, although they also occur commonly elsewhere. In these cases the form of development was influenced partly by detailed local factors such as ownership, location and physical resources, but overwhelmingly the shaping force was the needs of industry developing in a competitive laissez-faire environment. Little centralised planning occurred and the industrial city became dominated by its core, the needs of industry and the prevailing transport technology. As Goheen (1970: 11) explained for nineteenth-century Toronto 'industry was able to demand almost any land in the city, such was its bidding power and such was the utility which manufacturing gave to the land'.

The power of industry in shaping the city was extensive for two reasons. First, manufacturing industry became the prime motor of urban growth and the dominant, although not the largest, land user in the city. But its influence went much wider. Industry reshaped the social map of the city and played a large part in determining and providing the housing needs of the urban populations. In addition, the industrialists came to dominate the civic and other institutions which shaped the political and cultural life of the city. Even wider than this, the nations which first created large urban/industrial agglomerations, notably Britain, Germany and the USA, were stamping

4

their own imprint of development upon large parts of the rest of the world in the nineteenth and early twentieth centuries.

Second, and from a more purely academic perspective, the industrial/ urban form proved to be an enduring role model for development throughout the western world, certainly up to the mid point of the present century. Much of our understanding of the city, including the derivation of a large body of urban land theory which will be reviewed in Chapter 2, thus stems from it. Clearly, in the final quarter of the present century, the large industrial city with its traditional land use needs is undergoing profound change.

One of the major underlying questions of this book therefore is the extent to which our understanding of and planning for the city, based as they are upon a largely outdated model, are valid for present day and future needs.

The typical urban morphology which emerged in the latter phases of industrial growth in the mid twentieth century carried forward many of its earlier elements, but it also introduced important changes. These will be discussed more fully later, but they need to be introduced here. Above all, the morphological anarchy of the nineteenth century became tamed by the introduction of comprehensive town planning. This had many effects, amongst which the encouragement of hierarchical forms of urban centrality, the segregation of urban land uses and the introduction of a degree of order and tidiness were important. New forms of transport permitted a greater degree of specialisation in land use and created new locational imperatives. New building standards and codes were enacted as a response to the most squalid and haphazard aspects of earlier urban conditions. Social require- ments in the form of community facilities, better space standards and amenity and environmental considerations were brought to the fore. Building materials and techniques became more standardised so that a greater degree of uniformity became the norm. Huge swathes of low density housing developments, high rise office and apartment blocks, efficiently laid out industrial estates and modular schools and hospitals were replicated from one city to another. Above all, the residential suburbs, catering for car owning residents became the characteristic morphological addition to the twentieth century city.

LAND AS POWER

Traditionally, the ownership, or occupation, of land has conferred great economic and political power. Originally, this was especially true of the large agricultural estates, but the cities too proclaimed the power and wealth of the major land owners, be they individual aristocrats, the monarchy or bodies such as the church. One of the greatest social changes prompted by the Industrial Revolution in Britain, was the transfer of power from the small number of wealthy 'landed gentry' to, at first a small, and then later a growing, number of urban and industrial interests. From this point onwards the power and influence of the individual land owners declined rapidly, except for those few like the Howard de Waldens and the Dukes of Bedford, Westminster, and Portland who were fortunate enough, or shrewd enough to possess large holdings in the growing urban areas.

In the latter part of the twentieth century, land, especially in urban areas, has retained its allure as a source of wealth, power and status, but the owner-ship pattern has changed significantly. There are still a few individuals who have made large fortunes from urban land, but for the most part the twen-tieth century has seen the emergence and steady growth of three different kinds of land owning groups. The first of these is a set of institutions, including pension funds, insurance groups, investment companies and other corporate interests. These bodies are interested in urban land for a number of reasons. Most obviously, they may actually use land and buildings to accommodate their own office functions, and in this they may be looking for prestige and corporate identity as well as functional office space. Who could fail to notice for example the Woolworth or Pan Am buildings in Manhattan or the National Westminster Tower in the City of London. More signifi-cantly, however, urban land and property represents to such bodies a commodity in which wealth may be stored and which may be traded, often with good prospects for capital gain and favourable tax conditions.

The second major set of urban land owners, which has also grown markedly during this century, is located in the public sector. This includes central and local government, public utilities, state owned industries and a number of other state and quasi-state bodies. Their involvement in land ownership stems from two different reasons. First, a number of public bodies exist in order to provide services, most of which require land, e.g., for schools, hospitals, roads, and housing. Second, for reasons of social equity and planning efficiency, public sector control may be exerted over land develop-ment and urban expansion. In a number of European countries in particular there are long traditions of municipal land ownership, and state bodies are often involved in the process of land development.

The third group of owners is the growing army of individual house owners. These have been responsible for a greater fragmentation and dispersal of land holding than ever before. Their motives are twofold. Most obviously there is the status, social advantage and freedom which is conferred by home ownership, but there are also economic benefits in the form of favourable tax arrangements and the prospect of long term financial security.

There is another sense also in which land is associated with power and control. Drawing upon the work of Michel Foucault (1984), Cullen (1990) has argued that political treatises in the eighteenth century began to bracket architecture, land and space together with control and power. Foucault is interpreted as arguing that the city, its built form and the practices of those

6

who sculpt that form are largely concerned with imposing institutions and structures of control over individuals.

LAND AS THE BASIS OF THE PLANNING SYSTEM

The two major themes outlined above, land as urban morphology and land as power, come together to form a large part of the basis of town planning. In Britain and elsewhere, the planning profession which began to emerge at the beginning of the twentieth century was concerned with many aspects of urban development, but new and more pleasing forms of urban layout and a concern to protect the interests of the weaker groups in society were central. In both of these endeavours, the use, disposition, ownership of and access to land were key factors. In short, town planning was largely synonymous with land use planning.

As far as morphology was concerned, the central task was seen as overcoming the worst aspects of the congested, and insanitary jumble of land uses which had typified the industrial city. Attempts had been made by a few enlightened industrialists in nineteenth century Britain and Germany, to build model settlements in which the needs of industry and its workers could be satisfied harmoniously, for example, at Port Sunlight near Liverpool. Early planning documents often proposed utopian urban forms, some of which sought a return to more formal, pre-industrial layouts, whilst others attempted to deny urbanity by incorporating large amounts of open space and rural imagery. The interwar years were particularly influential, for they produced the International Congresses for Modern Architecture (CIAM). These paved the way for Le Corbusier's ideas on the Ville Radieuse with its tall blocks of flats, wide areas of public open space and careful segregation of motor vehicles and pedestrians.

Many of these ideas were taken up in practical form after the Second World War, notably by the British new town movement, but whereas other European countries were emphasising technical innovation, in Britain the focus was firmly upon the vernacular. The layout of the American city too was influenced by the modern movement but here it was given a particular character by Frank Lloyd Wright. This took development in a very different direction where the preference for individual family homes and gardens was to be combined with the freedom of the motor car in a virtually anti-urban form known as Broadacre City.

The impact of these elements of the modern movement upon the morphology and land use of today's cities has been considerable. Le Corbusier's ideas have been used, albeit in a penny-pinching form, in high density residential developments in cities throughout the world. The new town movement culminated in Britain's postwar development of more than thirty new towns, a style of urban development widely copied elsewhere and even more extensively reproduced in debased form through a multiplicity of

7

suburban housing estates. Lloyd Wright's ideas too shaped the development in North America, and increasingly in Europe, of low density cities with out-of-town shopping centres, leisure complexes and dispersed luxury housing developments.

Following the land as power argument takes us also to the heart of the planning system. In Britain especially, but also elsewhere, planning consists largely of the three elements identified by Peter Hall (1980). A professional bureaucracy forms the centre, and this is surrounded by a number of pressure groups and politicians. The first and third of these groups seek to balance and accommodate the demands of the middle group. The system is thus inherently political and it is significant that in seeking to analyse the politics of local planning, Blowers (1980) entitled his book *The Limits of Power*.

A large part of the justification for a planning system is that it resolves competing claims over the use of resources (especially land), attempts to balance an uneven distribution of power and protects the interests of weaker groups. In a practical sense, this includes the provision of land for community facilities and for housing poorer members of society and ensuring that some checks are kept upon the dominant land using activities. Much of the development of postwar planning in Britain, notably that in the new towns, placed great emphasis upon relatively egalitarian approaches and the importance of community values. Lord Reith (New Towns Committee 1946) was quite clear that these communities should be balanced, with a contribution from every type and class of person. The success of the planning system in protecting the weak, particularly in the provision of housing and community facilities, should not be discounted, but the 'Robin Hood' view of planning taking from the rich to give to the poor is not always appropriate. In the field of urban renewal, especially, there are abundant examples where relatively weak local communities have been pushed aside by a collusion of local authority and property development interests. In recent years, the pro-development policies adopted by a number of governments in order to combat urban decline have tended to downgrade the power of local and community interests.

Many examples of this can be seen, especially where the rise of an office economy has required a fundamental restructuring of urban land uses and a new economic infrastructure. In the USA the needs of urban renewal in some cities have resulted in powerful corporations persuading city governments to do much of the job of land assembly and provision of infrastructure (Friedland 1982). Following the 1949 Housing Act, city and federal governments began acting effectively as brokers for private developers. In the process they biased the power struggle between the political liberals and labour leaders who wanted low cost housing to replace slums and the corporate interests who wanted commercial and high rent residential development. The situation heralded what has happened in the London Dockland redevelopment in the 1980s. Such developments may well be desirable in the interests of the overall urban economy, especially where they follow a long period of decline or dereliction, but they lend weight to Cherry's question of 'Whose values are being protected, and by whom?' (1982: 86).

LAND AS ENVIRONMENT

During the 1980s, there has been a renewed concern over the quality and protection of the environment. Although much of the interest has been expressed at a global scale, or has involved 'natural' landscapes, there are also many local issues concerning urban environments. Much of the original motivation for the development of suburbs came from the desire of wealthy people to construct pleasant environments which were separate from the congested city, and similar motives help to explain counterurbanisation trends today. Within the city itself, environmental concern was first translated into exclusive parks and walkways for the nobility, but by the nineteenth century municipal parks for all citizens were becoming common. Much of the imagery of planning, be it for municipal parks, garden suburbs, green belts or the design of contemporary office developments, has invoked the idea of bringing the countryside into the city.

URBAN CHANGE

The past two decades have been a period of far reaching change for urban areas in the developed nations. Almost all urban activities have been affected by the processes of change, to such an extent that even the concepts of urbanisation itself must be questioned. Naturally, there are profound consequences for our understanding of urban land issues.

Perhaps the most basic factor to be confronted is that a number of powerful forces have been conspiring in those countries with market economies to direct growth away from large cities. By the middle of this century there were signs from some of the largest cities, including London, that after nearly two hundred years of rapid urban growth the rate was slackening. In the middle 1970s a reversal from urban population concentration to deconcentration was widely identified in the USA by Beale (1975), Berry (1976) and Vining and Strauss (1977) amongst others. The results of the 1981 and 1991 censuses revealed that the process was well established in Britain too, and in fact had been since the 1950s. Hall and Hay (1980) summarised the growth trends, particularly from North American evidence, as moving:

1   from larger to smaller metropolitan areas
2   from metropolitan cores to fringes
3   from urban to rural areas
4   from older manufacturing areas to newer service areas.

It soon became clear that these processes were very widespread and complex in their impact. The fact that they went beyond mere suburbanisation and local decentralisation led to the use of the term counterurbanisation. This process of counterurbanisation has been comprehensively reviewed for a large number of countries by Champion (1989). In this review, Frey (1989) suggested that it was the 1970s which represented the important decade of transition with counterurbanisation in the USA, at least, being a reaction to the economic shocks of the age. Champion concluded that by the end of the 1980s counterurbanisation was widespread but not universal. Some countries and cities continued to experience urban concentration and in others there were signs that the pace of counterurbanisation was slowing. Population figures for London, for example, suggest that some growth has resumed since 1984 and this prompts the possibility of a fresh reversal, back to the pre-turn-around pattern. Until further evidence becomes available there must remain some doubts about whether the 1970s represented a temporary hiccup in the pattern of urban growth, linked perhaps to cyclical economic depression, or signalled a long term tendency towards population dispersal and urban decline.

Whether temporary or permanent, certain explanations can be cited in this pattern of urban restructuring. First there is an economic dimension, expressed through changing patterns of urban employment. The large industrial cities of the western world have lost much of their industrial *raison d'être* in the face of competition from the newly industrialising countries in the Pacific rim and elsewhere. This has resulted in a dramatic decline in the number of industrial jobs in the urban heartlands of Britain, Germany, France and the USA. For example between 1971 and 1987 the following manufacturing job losses were recorded: London Local Labour Market Area, 602,000; Birmingham, 149,000; Glasgow, 110,000; Manchester, 103,000; Liverpool, 101,000 (Champion and Townsend 1990). The manufacturing plants which remain in these countries have been forced to change their modes of production, reduce and restructure their labour forces and choose new locations. At the same time, substantial growth has been recorded in service sector jobs; in Britain, for example, over 2.5 million service jobs were created in the period 1971–84 (ESRC 1989). Almost without exception these jobs had different labour and locational requirements from the manufacturing jobs which had previously sustained the cities, and hence they led to different land use needs.

In the social sphere too the 1970s and 1980s witnessed far reaching changes. The impact of changing age structures, family sizes and marriage patterns upon housing requirements and urban development has already been alluded to. In addition, changing lifestyle preferences, growing environmental awareness and increasing personal affluence for many encouraged people to seek work and residence in small towns with better living conditions, cleaner environments, lower crime levels and lower local government rates. In the process the size and power of the big cities has been weakened.

These two processes have been encouraged by improvements and growth in transport which has been both a prime mover and an enabling factor. In the West European countries for which statistics are readily available, traffic volume measured in millions of vehicle kilometres, increased on average by 17.5 per cent between 1984 and 1988. In the USA, starting from a much higher baseline of vehicle useage, the increase was 8.1 per cent. (International Road Federation 1989). In Great Britain, the ownership of cars and light goods vehicles increased by 23 per cent in the decade 1975–85 (Dept of Transport 1986). On the one hand transport congestion and delays in large cities have acted as a disincentive to industrial location, and on the other hand the growth in motor traffic has opened up outer suburban areas and small towns, both for industrial/office employment and for residential purposes. Even changes in the technology and handling of seaborne freight, and the dramatic decline in liner passenger traffic have hit many large port cities severely. This can be seen particularly clearly in the contemporary land use patterns of major dock areas.

Finally, government policies have played a significant part in the pattern of differential growth between large cities and other localities. In Britain, for example government policy during and after the Second World War actively encouraged the decentralisation of industry. Regional policy as variously operated in the postwar years, coupled with new town development programmes resulted in economic activity deserting the older and bigger cities in some measure. Similarly, in West Germany the regional planning policy (Kontuly and Vogelsang 1989) and in France the *villes moyennes* programme (Winchester and Ogden 1989) militated against big cities.

Alongside counterurbanisation, another major concept which helps to articulate the changing land use needs of modern urban society, is that of the postindustrial city. Hall (1988), in particular, suggested that Britain at least is well along the road to a postindustrial economy. Somewhat earlier he had suggested, along with others (Brotchie *et al.* 1985) that the microelectronic information technology revolution was beginning to produce changes in our pattern of living and working at least as profound as those produced by the Industrial Revolution, but within a shorter time frame.

The land use consequences of these trends, especially those of counter-urbanisation and postindustrialism are still unclear and subject to intense speculation. But the general implications for urban and quasi-urban land patterns can be outlined. In short, the massive forces of economic and social restructuring are having, and will continue to have, profound significance for urban land. The overall pattern of change can be broken for convenience into two major components.

On the one hand, most large cities show signs of serious inner area problems. The loss of job opportunities, the concentration of deprived social groups, the collapse of the urban infrastructure, the unpopularity of many public sector developments and the difficulty of attracting private sector investment have all combined to sweep away the previous vitality of many of these areas. The immediate consequence has been to throw the inner city land market into turmoil. Land which once had a relatively high demand, and hence value, attracts little commercial interest once those traditional uses have deserted the city. As a result, vacant and underused land has become a common characteristic of such areas.

Conversely, the demand for land on the urban periphery, and in small towns throughout the outer metropolitan fringe is relatively buoyant. In Britain, Hall (1988: 16) suggested that even if the major urban agglomerations succeed in stabilising their populations, the continued growth of what he terms the Golden Belt and Golden Horn regions in the southern half of the country will 'entail a voracious demand for conversion of rural land for urban purpose'. This then is the general context, a massive and relatively recent restructuring of urban areas and their economies, even to the extent that our traditional notions of the city as a settlement form must be re-examined. It raises many issues about our understanding and use of urban land, which will be addressed in the following chapters.

2

URBAN LAND ALLOCATION

In broad terms, the allocation of land, like the allocation of other commodities is determined by the nature of the politico-economic system in which it is set. In traditional societies this was normally a function of custom or convention (including religion), but in modern urban societies there are fundamentally two schemes. First, there is the market economy which began to dominate in industrial societies, starting with Britain, in the early eighteenth century. This is based upon individual ownership of land, not necessarily wholly private, with the wishes of buyers and sellers being brought into balance through the exchange and price mechanisms of the market. Second, there are variants of the command economy in which there is centralised decision making and a very high level of state intervention and ownership. In the broad political sense, such systems are usually modelled upon Marxist lines, but in a narrower sense elements of centralised control are present in most of what we understand by town planning. The purpose of this chapter is to look at the processes which shape the allocation and patterning of urban land according to each of these principles. It should immediately be pointed out that as we are dealing with developed western countries the distinction is not clearly polarised, but is one of a market allocation of land controlled by greater or lesser amounts of public intervention. At the outset, it is necessary to give some consideration to the particular nature of urban land itself, especially since this influences its treatment as an economic commodity.

THE PARTICULAR NATURE OF URBAN LAND

Land is unlike most other commodities involved in the production process because it possesses a number of unusual and complex characteristics, the most important of which are outlined below.

1   *Fixed supply.* In general, land is considered to be in fixed supply because no more can be created. There are, however, important qualifications to this: reclamation can add to the total stock, greater intensity of use can increase the effective supply and the amount available locally can be

13

increased if land owners bring more on to the market or if urban development is allowed to spread outwards onto agricultural or other land. A 1-kilometre extension to a city with a radius of 4 kilometres increases the area by over 50 per cent. The nature of urban development and the impact of planning creates, in effect, submarkets, and the supply of land in one category may be increased by a reduction in another.

2   *No cost of supply.* In an absolute sense, land can be considered a 'gift of nature' with no cost of creation, except in rare cases of the reclamation of land from the sea. In reality, of course, there are costs of providing infrastructure, development, improvement and other inputs to be considered.

3   *Unique/irreplaceable.* Each plot of land is unique in terms of size, configuration, physical characteristics and location. For these reasons no plot can be exactly replaced by another.

4   *Immobile.* Land is permanent and cannot be moved, although a limited degree of flexibility can be achieved through the substitution of transport costs.

5   *Permanence.* Land is uniquely permanent. It may be altered or damaged and it may be subject to the law of diminishing returns for a particular form of development, but in the urban context it is generally indestructible. The buildings erected upon it must be viewed rather differently, but even here there are very long lifespans to be considered and the inherited pattern of urban development is a powerful constraint.

As a result of these basic characteristics, together with the legal, social and political structures which different societies have developed, the use and ownership of land involves an enormously complex package of interests, rights and occupancy. Some of what happens to land depends upon the decisions and actions of the owners or occupiers, but much is also determined by the actions taken by adjacent owners and the broader society. Finally, a number of unpredictable, non-economic factors, including prestige, symbolism and social values need to be entered into the equation.

MARKET FORCES

Demand factors

For reasons connected with the particular nature of land, outlined above, it is normally assumed that demand factors are far more important than those of supply when considering the allocation of urban land. Supply is taken to be relatively inelastic, so it is essentially demand which sets the price.

Leaving aside the influence of public intervention, which will be discussed in a later section, urban land use in a market economy is determined by the decisions made by individual firms, households and other bodies with regard to jobs, housing, shopping and many other urban activities. Each of these requires land and the activity which can outbid all others will acquire the site. There is thus a presumption that land will always go to the most profitable use. Harvey (1987) suggested that a number of assumptions are built into the way in which the market handles these decisions, ie.

1   Resources are allocated on the basis of prices, costs and profits.
2   Firms and households will have locational preferences which are reflected in land prices and rents.
3   Owners sell and rent to the highest bidder.
4   Buyers and sellers have sufficient knowledge of the market to provide competition.
5   There is no government interference.
6   There are no dynamic changes in transport or technology.

From the point of view of the individual household or firm using urban land a number of general factors are important. Again, following Harvey, these may be summarised as follows.

1   *General accessibility* with the centre, or CBD, traditionally being considered as the most accessible point and the focus for transport, labour and retail markets.
2   *Special accessibility* as conferred by agglomeration economies such as common services, specialised labour supplies and complementarity for businesses or by social reputation and status for households.
3   Additional factors: including historical, topographical and other special site characteristics.

In practice, the bidding process by which firms and households seek to obtain land through the market thus takes account of an inseparable package of attributes, including location, proximity of services, facilities, complementary activities, neighbourhood quality, social factors and transport. This gives rise to the notion that land values depend not simply upon site characteristics *per se* but upon the wider actions of society, past, present and future.

But even this is to take an unduly simple view of the process by assuming that demand is determined by purchasers or renters who want to use the land directly and immediately. In fact demand has two components which need to be separated (Goodchild and Munton 1985: 1) demand from purchasers who want to use the land and whose main concern is its value derived from current rent or utility; and 2) demand from investors wishing to enjoy an increase in value derived from future expectations.

This reflects the distinction, emphasised in Marxist analysis, between the use value of land, that is, its utility for a particular purpose such as housing or industry from which a financial or other benefit can be obtained, and the exchange or sale value which is determined by economic and social transactions both currently and in the future. With the growing strength and activity of financial institutions such as banks, insurance companies, pension funds and development companies, together with the relative security of property as an investment, the importance of exchange value has tended to be high in recent years. Urban sites are commodities which are traded and where the value can be crudely determined using formulae like that suggested by Wendt (1957):

$$\text{Land value} = \frac{\text{(aggregate gross revenues) minus (expected costs)}}{\text{Capitalisation rate}}$$

Amongst the factors which affect the revenue are the size and activity level of the market, income spent on services, the competitive pull of the particular urban market and likely public investment in improvements. Costs are affected by local taxes, operating revenue, interest rates and depreciation. Capitalisation is affected by interest rates, allowances for risk and expectations of capital gain. In Los Angeles, Titman (1985) examined a number of vacant and undeveloped plots and concluded that speculative holding for more valuable development in the future was the key to the pattern. Similarly, in a simple model, Capozza and Helsley (1989) suggested that a growth premium, in the form of expected future rent increases may account for as much as a half of the average price of land in a rapidly growing city. This brings us to the supply side of the market.

Supply factors

Given the assumed inelasticity of supply and the other factors listed at the beginning of the chapter, it is perhaps not surprising that the supply side of the urban land market has been neglected. Certainly, we can accept that the supply side may be seriously constrained, but no longer can it be argued that it is fixed in amount and incurs no supply or production costs.

The suggestion that the supply side of the market needs to be considered more actively has been gaining strength in recent years (Evans 1983; Wiltshaw 1985; Goodchild and Munton 1985). The argument can be summarised as follows. Three features need to be considered. First, planning has an effect upon supply through the way in which decisions are taken to permit or deny development. Second, there are physical constraints in the form of land quality or the presence of a major barrier which may limit urban activity. In a study of forty-five cities in the USA, Rose (1989) concluded that planning restrictions and physical restrictions (especially stretches of water) accounted for 40 per cent of the observed differences in land prices. Third, and more elaborately, are the behavioural choices made by land owners. As will be seen in Chapters 6 and 7, the willingness of land owners to be involved in the development process is a powerful influence upon the subsequent course of events. The strongest interpretation of the land owner's behaviour involves monopoly power, because each site is unique and the land owner can determine the price by withholding it from sale (Neutze

1973; Drewett 1973). This notion of monopoly is a powerful one, widely discussed by Marxist commentators, but the practical reality of the urban land market is that although some sites may attract a price premium, very few achieve monopoly.

In arguing that land does have a supply price, some account must be taken of how this is set. The land owner's price may be based upon comparisons with other nearby sites or the cost of replacing his land with an alternative site, but it may also involve an erroneous reading of the market, uncertainty about future price movements, or simply greed (Pearce *et al.* 1978). In considering when, or whether, to sell, the vendor also takes into account the cost of moving, the loss of amenity and a range of other nonfinancial benefits attached to land ownership (Popper 1978; Neutze 1987). All of this means that the supply price must exceed the existing value if the owner is to be induced to sell. Here selling must be distinguished from renting because the former is a permanent transaction whereas the terms of renting can be varied or rescinded. The behaviour of land owners, or suppliers, in aggregate determines the overall supply of land, and in certain circumstances it is possible to see that this is far from inelastic. The total amount of land coming on to the market can vary substantially over short periods as the cyclic market for housing land in Britain shows. For example, in 1970 land for 70,000 houses was sold, but in 1973, when land values had tripled, land for 170,000 houses was supplied (Neuburger and Nichol 1976). Finally, the supply price may also be affected by local tax conditions, with suppliers adding development taxes to their selling prices, and by the costs of clearance and reclamation involved in bringing derelict land into a marketable state.

In conclusion, a mixture of interacting influences can be seen to determine the way in which the market allocates urban land. Neoclassical economic views have emphasised the profitability and utility of competing uses, as mediated through accessibility and rent levels. Despite this oversimplification of a very complex process, it has tended to dominate the derivation of models of urban land allocation and use. In the next section some of these models will be examined in more detail.

MODELS OF URBAN LAND USE AND LAND ALLOCATION

Many models of urban land have been developed by economists, geographers and others and a large proportion of them can be described as bid-rent models. These assume that land using activities have different needs to locate close to the centre of the city and will bid for land accordingly. This results in a gradient of intensities of land uses and land prices which declines outwards from the centre in a more or less predictable manner. All activities are thus optimally located, such that utility, or profit, is maximised. The provenance

17

of most of these models can be traced to the work on agricultural land published by Von Thunen in 1826, although it is Hurd (1903) who is usually given credit for applying it to urban areas.

The further development of these models can be seen through the works of Isard (1956), Beckman (1957), Wingo (1961), Alonso (1964), Muth (1969), Mills (1972) and Miyao (1981).They have been well summarised by Balchin and Kieve (1977), Hallett (1978) and Hudson and Rhind (1980), so a brief outline will suffice here. It should also be noted that, more recently, Fujita (1989) has attempted to extend them beyond the limited explanation of positivist theory into normative theory with the identification of efficient spatial structures and means of achieving them.

In essence, the bid-rent formulations rest upon the assumption that different activities will have bid-rent curves which vary in form according to their need to be at the centre of the city. This, in turn, depends upon the nature of the activities, their ability to take advantage of highly priced central sites and their sensitivity to transport costs. A number of commercial activities, for example, have very specific labour needs, customer requirements and linkages with other activities. All of these can, theoretically, best be satisfied at the centre where transport facilities maximise labour availability, customer flow and proximate linkages. Thus they will be prepared to pay high prices and will have a steep rent gradient. A number of industrial activities have (and had even more strongly in the past) a need to be close to the centre for reasons of labour availability, transport services and marketing services, but their need is less than that for commercial uses and they are less sensitive to small variations in accessibility, therefore their rent gradient is less steep and they cannot compete successfully for the very central sites. Residential activities are normally the largest user of land in the city. They may desire a fairly central location (although suburban qualities are increasingly preferred), but they cannot derive sufficient utility or profit to outbid commerce and industry. In effect, they become a residual use, consigned to the lowest levels of the bid-rent curve with locations furthest from the centre. This theory provides the rationale for the arrangement of land uses and values indicated in Figure 2.1.

An overall land value surface can be seen in Figure 2.1, and if the diagram is rotated about its vertical axis at the city centre, a broadly concentric zonation of land use is achieved. The overriding fact from this is that land use is seen to determine land value. Point A marks the distance from the centre where the decline in interest from commercial activities is such that industrial uses can outbid them and thus become the dominant activity. At point B, similarly, residential uses compete successfully with industry. From the centre to point A commerce dominates, but industry, and at a lower level even housing, would be subordinate uses.

For all activities there will be a trade-off between the high costs of central area land and the high costs of transport incurred by locating further out, but

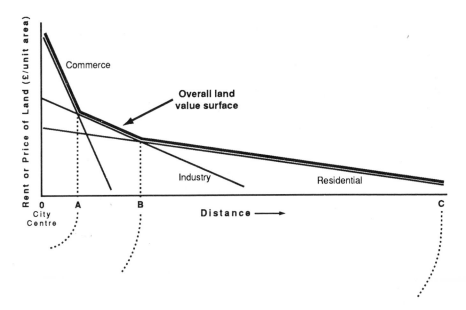

Figure 2.1 Urban land uses and the bid-rent model

the effects of this may be more apparent for residential uses, or for individual households within the residential sector, for whom the utility of a central location is lower – Figure 2.2.

These notions of bid-rent theory, and the pattern of land use which is assumed to result, provide a degree of explanation for one of the best known models of urban structure, that of E.W. Burgess (1925). It must however be stressed that Burgess did not use rent theory as such. He was a sociologist and his model was derived from empirical observations of the way in which the city of Chicago had developed. As such, it is a hybrid of idealised land use patterns and urban social structure with a strong emphasis upon residential areas. The model is commonly represented as a purely concentric zonation of activities, but it is important to remember that, as originally developed from Chicago, it had significant additions in the form of specialised sectors – Figure 2.3.

The importance of a sectoral form of development was taken up by Hoyt (1939) in his study of residential rent levels in a large number of American cities, including Chicago. Although there are obvious difference between these two models there are also some similarities – Figure 2.3. The main difference is that Hoyt considered direction, as well as distance, from the CBD to be important in determining land use. The arrangement of the sectors was such that high income areas were protected from low income districts and from industry by buffer zones of middle income housing. The strength of the sectoral pattern which emerges should not be allowed to

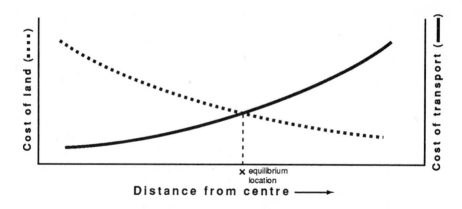

Figure 2.2 Variation in land and transport costs with distance from city centre

disguise the fact that within the sectors, Hoyt clearly identified concentric zones of differential rent and a tendency for the most fashionable residential areas to migrate outwards from the centre along specific sectors. The processes differ, in that Burgess was largely concerned with social factors such as ecological competition and migration, whereas Hoyt concentrated upon amenity value and filtering, but there is a case for suggesting that Hoyt's model should be considered as a refinement of that of Burgess rather than something which strikes off in a totally different direction. In seeking general applicability, Hoyt's model is to be preferred to that of Burgess because it is more firmly based in empirical data, being a synthesis of twenty-five cities, and because it goes further towards acknowledging that the CBD is not the only commercial focus of the city. This latter point is important because some of the rationale for identifying sectors was the growing import-ance of motor transport which was creating subsidiary commercial nodes, often alongside major radial roads, in the interwar period.

In a third major modelling exercise, Harris and Ullman (1945) took Hoyt's subtle recognition that the CBD was not the only focus of activity, and made it explicit in their multiple nucleii model – Figure 2.3. Again, the model is to some extent evolutionary in that it incorporates elements of Burgess and Hoyt, but it is more flexible than either. Essentially, it implies that the city has a cellular structure within which a number of specialised areas develop. Some of these may be highly nucleated, such as suburban shopping centres, but others may be quite large districts dominated by a single land use such as industry or upper-class housing. In relaxing the dominance of a single centre, and the assumptions of general accessibility to the core, the multiple nucleii model recognises the interaction of a number of locational factors. First, it gives some weight to topographical and historical features in the origins of the growth of the city, for example, in the absorp-tion of minor settlements. Second, it recognises that different levels of

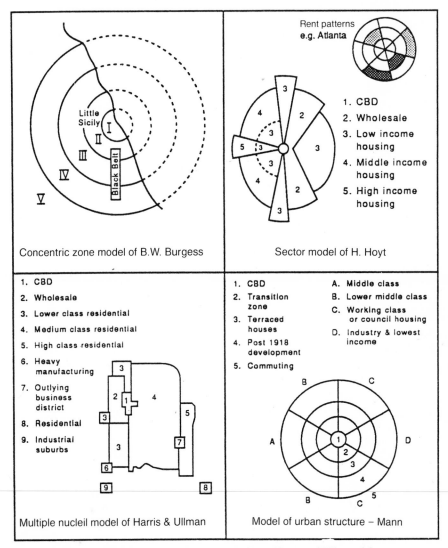

Concentric zone model of B.W. Burgess

Rent patterns e.g. Atlanta

1. CBD
2. Wholesale
3. Low income housing
4. Middle income housing
5. High income housing

Sector model of H. Hoyt

1. CBD
2. Wholesale
3. Lower class residential
4. Medium class residential
5. High class residential
6. Heavy manufacturing
7. Outlying business district
8. Residential
9. Industrial suburbs

Multiple nucleil model of Harris & Ullman

1. CBD
2. Transition zone
3. Terraced houses
4. Post 1918 development
5. Commuting

A. Middle class
B. Lower middle class
C. Working class or council housing
D. Industry & lowest income

Model of urban structure – Mann

Source: Original drawing based on Burgess, Hoyt, Harris and Ullman, Mann

Figure 2.3 Diagrammatic models of urban structure

retailing do not all seek a central site, some preferring suburban locations closer to their market. Third, it allows for the agglomeration economies and both the negative and positive externalities which cause certain firms or households to cluster together. These three models have become so well established in the literature on urban structure that they are normally referred to as 'the classical models'. They are all, quite clearly, North Amer-

21

ican in origin but an attempt to refine them for the British context has been made by Mann (1965). His diagrammatic model – Figure 2.3 – draws most heavily upon the concentric zone and sector models, but it makes passing reference to separate and more specialised areas. Importantly, it also makes allowance for public sector intervention in the form of local authority housing.

Naturally, models which have been around for such a long time will have attracted a wide measure of criticism (Carter 1976; Hallett 1978; Hudson and Rhind 1980), and the evolution of urban land use in the past three decades highlights some of their deficiencies. Some of these criticisms are valid, but in a sense unfair. For example, both Burgess and Hoyt were concerned with the rapidly growing American city of the first third of the twentieth century; to apply their models outside of this geographical and historical setting is misleading. Burgess, in particular, claimed no wide applicability for his model. Some attempt to deal with criticisms of the earlier models can be seen in the development of the later ones, notably the attempt to move away from the assumption of a single, overwhelmingly dominant central area. Similarly, there is the question of public intervention in the urban structure. At the time when Burgess was writing, the structure of the American city was almost wholly determined by market forces. Subsequently, and especially in European cities, state intervention through planning regulations, transport policy and the provision of public sector facilities has profoundly influenced urban structure. All of these models then have severe limitations. Burgess and Hoyt, in particular, simply described the patterns which they saw; they did not provide quantifiable models and they were not very explicit in their analysis of process. For all of their shortcomings, however, the models have an enduring quality and they have undoubtedly been fruitful in shaping our understanding of patterns of urban land use and structure.

These classical models have been attempts in varying degrees to provide comprehensive statements of urban structure. If we return now to the bid-rent models, it is clear that they mainly deal with more partial views. Residential land use has been their principal focus, and a number have looked at retailing patterns, but few have examined manufacturing or other activities. Putting the various urban subsystems together has generally not been attempted except in the simulation models such as that applied by Lowry (1964) to Pittsburgh.

The seminal work of Wingo (1961), Alonso (1964) and Muth (1969) looked mainly at residential choice and this has recently been extended by Thrall (1987) in a consumption theory of land rent. He made the usual assumptions of bid-rent models, including a city populated by identical households, on an isotropic plain with all employment at the centre. Households can decide about the consumption of two goods: land and a composite of all other goods. He argued that because of higher transport costs at the periphery, disposable incomes are lower, but to compensate for this land costs are lower

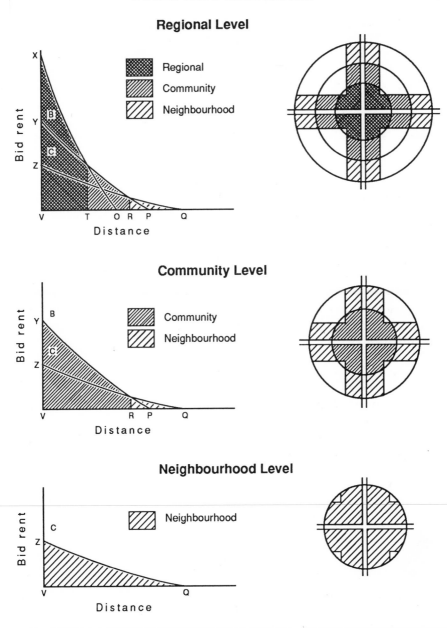

Source: After Garner 1966

Figure 2.4 The internal structure of regional, community and neighbourhood level shopping centres

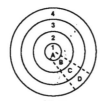

Nucleated Characteristics

Shop Types	Example Clusters
1. Central area	A. Apparel shops
2. Regional centres	B. Variety shops
3. Community centres	C.. Gift shops
4. Neighbourhood centres	D. Food shops

Nucleated Characteristics

Shops which represent in embryonic form the functional provisions of central area, regional, community and neighbourhood centres are arranged in relative order of threshhold values around the most nodal position of peak land values. Certain clusters of retailing specialisation may emerge, examples of which are furnished. The order of arrangement of shop types is not continuous in the real world, for there are clearly overlaps and variations in road patterns distort the nature of concentric belts.

Ribbon Characteristics

Shop Types	Example Clusters
1. Traditional Street	E. Banking
2. Arterial Ribbon	F. Cafes
3. Suburban Ribbon	G. Garages

Ribbon Characteristics

Shops which represent in embryonic form the functional provisions of outlying ribbon developments are more conspicuous toward the periphery of the CBD core and extend into and through the frame area. Certain clusters of retail and allied service specialisations may again emerge, and some of these create larger areas of functional specialisation in conjunction with other activities within the frame. The directional bias of these facilities is along the major thoroughfares.

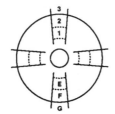

Special Area Characteristics

Shop Types	Example Clusters
1. High quality	H. Entertainments
2. Medium quality	J. Market
3. Low quality	K. Furniture
	L. Appliances

Special Area Characteristics

Shops which represent in embryonic form a response to special resource factors in terms of accessibility conditions may be differentiated in further functional clusterings of similar activities or alternatively according to quality levels. Clusterings of retail specialisations are in this case more distinctive as agglomerations of several associated functions, such as retail markets or entertainment districts, or particularly large space users, such as furniture marts and groupings of domestic appliance outlets. In location, these clusterings may take on characteristics of either nucleated or ribbon facilities. High quality shops may be found in concentrated positions anywhere in the central area, although more usually near to the peak land value node. Medium quality shops spread out over the greatest area of the CBD. Low quality shops seem to be most prevalent on the periphery and are usually equated with neighbourhood or arterial kinds of shopping provisions.

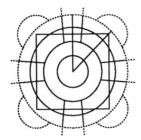

THE COMPLEX MODEL

This is by no means a stereotype of central area facilities , but indicates how component subsystems of the outlying retail pattern become combined together. The dashed boundary lines indicate the CBD core merges often imperceptibly with the frame, wherein certain other distinctive retailing characteristics may be found, especially in terms of large, conspicuous specialised functional areas. The most confused and complicated retailing parts are found towards the periphery of the core.

Source: Davies 1972

Figure 2.5 Structural model of Central Area Core Retail Facilities

and people consume more of it. At the centre higher disposable incomes allow families to consume more composite goods. Some doubts must be shed upon the reality of this because wealthy suburbanites and commuters do not have to trade off consumption of land against composite goods or transport costs; they can afford more of all.

The key concepts of the bid-rent model, i.e. land values and accessibility to the centre, have been applied in a variety of circumstances to the more specialised study of urban retailing, especially with respect to the structure of the urban core. The work of Garner (1966) and Berry (1967) drew attention to the connection between different levels of the shopping hierarchy and distance from the core such that a concentric arrangement of retailing land uses occurs – Figure 2.4. The idea was further developed by Davies (1972) for the city of Coventry and here we see (Figure 2.5) once again the combination of sectors and concentric rings. A detailed connection between shop types, land rents and distance from the centre, in the mainstream tradition of bid-rent theory has been outlined by Scott (1970) (Figure 2.6).

Implicit in all of these models of urban land use is the important role of transport, and it is necessary to consider two aspects. First, some reference must be made to the large scale simulation models of urban land use and transport which were developed in the united States from 1960 onwards, even though they are not primarily economic models. Second, it is useful to look at more detailed empirical examinations of the connection between transport and land costs.

The development of transport modelling created some of the prerequisites for land use modelling, (Harris 1985) and a number of massive studies of land use and transport have been undertaken, e.g. the Penn–Jersey Study and the Chicago Area Transport Study. Most of these models can be categorised as either:

1   non optimum-seeking models of disaggregated land use, e.g. Lowry (1964); or,

2   optimum-seeking models, often of a linear programme type, e.g. Herbert and Stevens (1960).

Early models took transport costs as given, but more recent ones take account of congestion, which means that travel costs are not uniform. A useful review of many of these model types was provided by Berechman and Gordon (1986). It is worth stressing that most of these models are 'non-economic' simulation models, i.e. they use rules taken from observed statistical regularities in travel and the location of activities, rather than from economic forces such as land rents and consumer preferences. Some attempt to reconcile the two categories has been made by Anas (1986). The integration of transport and land use models was explored by Putman (1983) who emphasised both the importance of transport for land use and the complexity of the relationship.

25

Source: Scott 1970; Kivell and Shaw in Dawson

Figure 2.6 Rent gradient in an unplanned shopping centre

Relatively few studies have empirically tested the relationship between land use and transport, but there are a number which have shown measurable connections between transport changes and land prices. In Britain, for example, improvements in urban rail services were shown to have increased house prices in Glasgow and London (Wacher 1971), and in the Tyne and Wear area (Pickett and Perrett 1984). Similar changes were reported following the opening of the Spadina subway line in Toronto (Bajic 1983). In Australia, property prices responded to the increase in petrol prices in 1978 (Evans and Beed 1986). Up until 1978, the value of houses in the outer suburbs of Melbourne increased more rapidly than those in the inner suburbs – Figure 2.7 – in line with experience in most western cities. From 1978 to 1981 however there was a reversal of this pattern – Figure 2.8. Values in the inner suburbs rose by 30–50 per cent, whereas those in the outer suburbs remained static. In the USA the increases in oil prices between 1973 and 1979 also caused some adjustments, in that there was a trend towards smaller cars and a continued movement of people and industry towards the suburbs in order to decrease the journey to work distances. In Japan, on the other hand, it appears that rising transport costs have had little effect upon urban structure. Rapidly rising real incomes and the willingness of employers to subsidise commuting costs mean that urban residents can, to some extent, shrug off rising energy costs. Industry has not moved out of the cities as much as it has elsewhere because of its continuing strong links to port facilities (Getis and Ishimizn 1986).

26

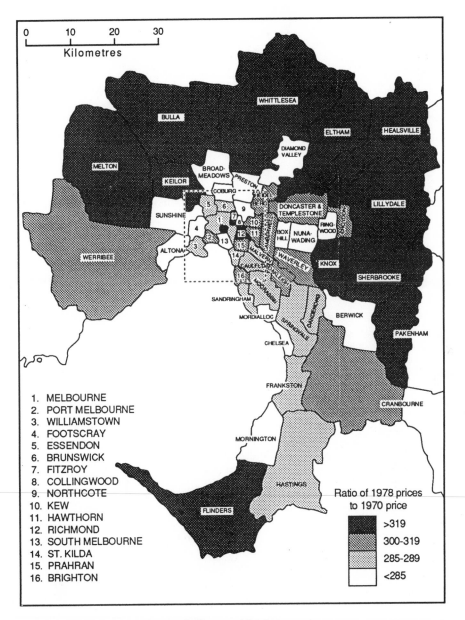

1. MELBOURNE
2. PORT MELBOURNE
3. WILLIAMSTOWN
4. FOOTSCRAY
5. ESSENDON
6. BRUNSWICK
7. FITZROY
8. COLLINGWOOD
9. NORTHCOTE
10. KEW
11. HAWTHORN
12. RICHMOND
13. SOUTH MELBOURNE
14. ST. KILDA
15. PRAHRAN
16. BRIGHTON

Ratio of 1978 prices
to 1970 price

>319
300–319
285–289
<285

Source: Evans and Beed 1986

Figure 2.7 Changes in house prices, Melbourne, 1970–8

Criticisms of economic models

A number of criticisms of individual models have already been noted, but it is also necessary to take a broader view. For simplicity, three headings will suffice – oversimplification, market failings and Marxist critiques – but, of course, there is also some overlap between these. It should not be assumed that these criticisms are always negative and destructive of the models, for in a number of cases they point the way towards better, albeit partial, understanding.

Oversimplification

In order to be meaningful, a model must be reasonably simple, but there are dangers of oversimplification. Most of the models referred to above ignore the physical setting of the city together with its inherited stock of land uses, they assume independence between land uses, they take the centre to be the point of maximum accessibility and the concentration of all employment and they assume a perfect market in which people behave in an economically rational manner deciding where to live solely on the basis of transport and land costs. Perhaps it is no wonder that Hallett (1978: 16) described the bid-rent models as 'a grotesque simplification of reality'.

From a geographer's viewpoint, one of the most seriously misleading assumptions is that regarding the role of the centre of the city. There are two aspects to consider. One is the unreality of assuming a single, totally dominant centre, when most cities clearly have a number of separate, perhaps specialised focal points. The second aspect concerns the changing role of the centre of the city and its decline under the impact of decentralisation and counterurbanisation. The CBD itself may still be relatively prosperous but the decline of jobs, the decentralisation of many activities and the deterioration of the urban environment have all challenged the traditional role of the central city in the wider metropolitan context. In a number of British and American cities these processes have emphasised Burgess's zone in transition. An inner city area has resulted in which the enduring problem of derelict and vacant land is evidence that demand and land prices have fallen, in some cases temporarily, to zero. Comprehensive information on land prices is not available in the UK, but robust common sense suggests that the land value surface of the declining industrial city has changed in the way indicated in Figure 2.9, with a major dip representing these depressed inner city values followed by a marked rise towards the suburbs. The severity of the fall in value at the edge of the city will depend upon the extent to which the local planning regime permits or prevents urban extension into the surrounding countryside. The declining importance of the centre as the hub about which the urban land use pattern is organised has been further demonstrated by McDonald (1984) who noted in his work on Chicago that

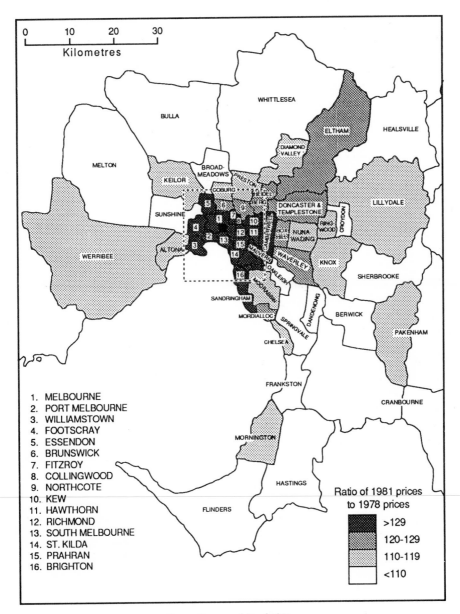

1. MELBOURNE
2. PORT MELBOURNE
3. WILLIAMSTOWN
4. FOOTSCRAY
5. ESSENDON
6. BRUNSWICK
7. FITZROY
8. COLLINGWOOD
9. NORTHCOTE
10. KEW
11. HAWTHORN
12. RICHMOND
13. SOUTH MELBOURNE
14. ST. KILDA
15. PRAHRAN
16. BRIGHTON

Ratio of 1981 prices
to 1978 prices

>129
120–129
110–119
<110

Source: Evans and Beed 1986

Figure 2.8 Changes in house prices, Melbourne, 1978–81

no theoretical model explained the variation in the land use in relation to distance from the CBD. In Los Angeles too, the CBD has been shown to have relatively little importance in determining land use patterns, especially within the residential sector (Heikkila *et al.* 1989)

Market failings

Even without involving the fundamental Marxist critique (which will be outlined below) it can be seen that the market in urban land and property has a number of shortcomings. One of the most comprehensive criticisms concerns its failure to cater adequately for a number of social needs, in terms of the provision of land for socially desirable but unprofitable uses, and in terms of catering for economically weak social groups. But in addition to this it has a number of more detailed, intrinsic shortcomings. Balchin and Kieve (1977: Ch. 2) describe the property market as 'one of the least efficient' and Ratcliffe (1976: 10) reports that it has been described as 'chaotic, monopolistic and irrational'. There seems to be some agreement about its shortcomings and these include: the uniqueness of each site, the imperfect knowledge of both buyers and sellers, the varying motives and amounts of power of the different participants, the expense and legal complexity of transfers, the length of legal rights and property interests, the role of non-monetary factors such as sentiment, symbolism and pride of ownership, the difficulty of assessing prices in a thin market, the importance of inertia and the longevity of individual buildings. In view of all these shortcomings it is no surprise that the perfect equilibrium between supply and demand which the market theoretically provides is also something of a myth. In reality, many sites are used sub-optimally and the process of adjustment to changing conditions is quite slow. It is relatively easy to list shortcomings of the market in this fashion, but it is important also to pose the question of whether alternative systems are better at allocating urban land.

Marxist criticisms

Notions of rent, capital and land ownership were central to the writings of Karl Marx, and although he was working in a broadly agricultural setting, a number of writers have subsequently extended his ideas into the urban context. Clearly, some caution is necessary here, for whereas agricultural land is an independent production unit, where rent is set according to the plot's own characteristics, land in the city has its usefulness and rent largely determined by its linkages with, and access to, other land, buildings and urban facilities.

The starting point for most contemporary Marxist challenges to existing land use models and the operation of the market is the assumption of very great, possibly monopoly, power in the hands of a few land owners.

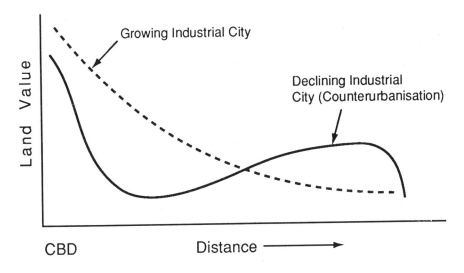

Figure 2.9 The changing urban land value surface

This enables them to manipulate or control the land market by charging absolute or monopoly rent. Harvey (1973) thus argued that in the CBD land prices do not normally depend upon locational advantage, but vice versa. Put in its extreme form, it is argued that the CBD exists not because of a genuine demand for its services, but because the land owning classes manipulate society's 'wants' to achieve high rents. Similarly, Ball (1985) suggested that simple demand is an insufficient explanation for the dense concentration of offices at the city centre. Instead, he favoured the explanation that the big investors, including pension and insurance funds, go for the development of office blocks in 'blue chip' locations in the city centre to minimise the risk of premature obsolescence. The process becomes self-reinforcing as others seek the same safe locations. Similarly, Marxists would argue that high house prices are due to high land prices rather than the liberal explanation that high demand for houses pushes up residential land prices. In practice, the argument of monopoly powers being used to deter-mine prices is difficult to sustain in the market for new houses, because at any given time the new housing available forms only a small proportion of the total stock for sale, the great majority of which is owned by householders acting individually. Harvey went on to develop the distinction between use value and exchange value, and demonstrated that existing models treat one or the other of these, but not both together. For example, the model of Burgess, and indeed most of the contributions made by geographers, sociol-ogists and town planners, deal with use value, whereas bid-rent models, and land economists in general, are more concerned with exchange values. He held out the possibility of a Marxist approach which would bring the two

31

concepts together, but such a model is elusive. For Harvey, rent was not simply the key to the land use but also a prime cause of spatial conflict, production inefficiencies and social injustice.

Rent is clearly central to the Marxist interpretation, but landed capital can also use other methods to control development in order to maximise its own interests (Edel 1976). These include: 1) large estates constraining suburban growth; 2) monopoly control over building land; 3) fragmentation of ownership; 4) deliberate delays to force up prices; 5) speculative holding of land; and 6) capture of state planning controls. Clearly, there is considerable scope here for influence to be exerted over both temporal and spatial aspects of land development.

Two final comments will help to bridge the gap between the powerful, but essentially detached, criticism of Marxist theory, and the reality of land allocation processes in practice. One is that an extensive study, largely sympathetic to the Marxist view, concluded that, in Britain at least, there is no single land owning group which could be identified as a coherent faction with anything approaching monopoly powers (Massey and Catalano 1978). The second is the observation that in the quasi-Marxist societies of Eastern Europe, after several decades of absence, the benefits of a market with land pricing and wider private ownership are tentatively being re-introduced.

PUBLIC INTERVENTION

Alongside the role of the market in allocating urban land, it is also necessary to consider the impact of a wide range of public intervention measures. These vary according to the political context in which the city is set. History reveals a large number of isolated instances of regulation in Europe by monarchical or other public bodies. These include the early code of Roman law which provided a number of building regulations, the medieval controls relating to protection from fire and nuisance, the grandiose town planning layouts of the sixteenth, seventeenth and eighteenth centuries and the late-nineteenth-century housing and sanitation by-laws in Britain. Comprehensive public intervention, however, is largely a feature of the twentieth century and it has gained in scope and sophistication as the century has progressed. Today, public control in matters of urban land use is generally more elaborate and far reaching in European countries than it is in North America and Australia, but even in the latter countries where individual rights in economic terms are highly prized, the unfettered pursuit of economic gain at the expense of others is not possible. Government laws, statutes and provisions are more far reaching with respect to land than any other commodity (Goldberg and Chinloy 1984). Public intervention is closely related to political ideology and in postwar Britain it has risen or fallen according to which political party has been in power.

Public intervention in matters of urban land use comes about for many reasons, but basically they relate to failures, or presumed failures, of the market. Many of these failings have been reviewed above, but the most important ones in prompting intervention are:

1   The fact that land comes on to the market in a haphazard and unpredictable way and at no time is it in equilibrium with all land being used optimally.
2   Because of the market's emphasis upon the price mechanism, it is not good at providing for public or merit goods. Socially necessary uses such as schools, hospitals or parks thus get squeezed out.
3   Strong land owning and financial cabals can dominate weaker groups.
4   The market is not good at dealing with negative externalities such as traffic congestion, noise and noxious industries.

Public bodies thus intervene in order to overcome these problems. A wide range of instruments is available, but for simplicity they will be discussed under three broad headings: public ownership, regulatory measures and fiscal measures.

Public ownership

Public ownership of land is closely dependent upon political ideology and it may vary from comprehensive state ownership (nationalisation) through to selective, and perhaps temporary, state or municipal involvement. Leaving aside the centrally planned economies of Eastern Europe, it is possible to trace a long history of public land ownership in Western Europe, but it is far less common in North America. The issue will be discussed more fully in Chapter 5, but a brief introduction is necessary here.

Perhaps the least controversial aspect of public land ownership is the acquisition of land for the provision of various public services such as roads, schools, hospitals parks or military establishments. Even in the relatively unfettered market conditions of the USA, this process operates, and there is also a right of 'eminent domain' whereby the federal or state government may acquire land compulsorily.

A higher level of public intervention in land ownership occurs in those countries where government or municipal bodies are heavily involved in the land development process. In Britain, public ownership of development land was an intrinsic part of the 1947 Town and Country Planning Act. It did not remain on the statute books, however, and two subsequent attempts to institute it were also repealed, the most recent being the Community Land Act of 1975, which was repealed in 1980. Only in Wales does the system remain, through the work of the Land Authority for Wales which buys, assembles and sells land for development. Selling development land for its full market value allows the authority to finance its future activities. Elsewhere in Britain, limited versions of the process also operate in certain run down inner city and industrial areas, where selective programmes such as the Urban Development Corporations, Derelict Land Grants and Partnership schemes empower public sector bodies to acquire land, assemble sites, reclaim and service land and bring it forward for development, especially by the private sector. These are not really examples of public land ownership *per se* but attempts to 'kickstart' the market.

Outside of Britain, many other European countries operate systems of public land ownership for development land. In German cities, for example, there is a relatively harmonious relationship between the free market and government intervention (Williams 1988) and both municipal and provincial governments maintain land banks for future urban development. In Scandinavia and The Netherlands municipal authorities are extensively involved in land ownership, often acting as a middle man between farmers and developers. In Stockholm a widespread system of municipal leasehold has long been in use, and as a result the city now owns three-quarters of the land within its boundary (Ratzka 1980). In almost all such cases, public ownership is combined with other fiscal or planning measures.

Regulatory measures

Even in the broadly market economies of the west, the state has a wide array of instruments for regulating urban land use. These range from the detailed application of such specific devices as building codes and safety regulations, to the full gamut of comprehensive town planning legislation. In general, regulation stems from an attempt to reconcile two of the state's contradictory functions. These are, accumulation, i.e. the creation of favourable conditions for capital growth and investment, and legitimation, i.e. the promotion of equity and social harmony.

As practised in Britain, and most other western nations, town planning is firmly rooted in land use planning. Healey *et al.* (1988: 17) suggested that 'the planning system both expresses and embodies some of the key rules of access through which agents may realise their interests in land, property, location, place and environment'. They identified development control, that is the state's power to approve or deny development, as the central focus. It is translated into a spatial pattern of land use control through the operation of district, structure and other plans which outline whether, and what kind of, development will be approved in particular locations. Successive large scale acts, such as the Town and Country Planning Acts of 1947 and 1971, and the Local Government, Planning and Land Act of 1980 have given British local authorities wide powers to control overall urban development. In addition, a number of more specific acts, for example the New Towns Act of 1946, and a succession of housing acts have had a important influence upon land use within their own areas of concern.

The planning system established in 1947 was designed largely as a control over land uses, especially the prevention of urban sprawl and the regulation of land uses within cities. Since then, there have been many changes, especially in the 1980s, but the fundamentals, including the definition of development and the system of development control, together with the relationship between day to day control and the broad development plans, remain broadly the same. In effect, the right to develop land is nationalised, and decisions about development are taken in the public interest quite separate from issues concerning land values or ownership. Essentially, it is a permissive system, in which the planning authority can approve or refuse permission to develop, but does not, in general, initiate development itself. Blowers (1986) drew attention to the paradox of planning being focused upon land use, over which it has little control (because a developer may fail to come forward, or because refusal of development permission may be overturned on appeal), but neglecting land values over which it does have influence by indicating both broad and specific locations approved for development. After major reorganisation of local government in the 1970s, which gave a great deal of planning power to county authorities and made the structure plan the most important document, there is now some indication of another change which will return key planning powers to the more local, district council level. A number of the planning related issues will be dealt with further in Chapter 6, but the reader is also referred to Healey *et al.* (1988).

In other West European countries there are a few similarities with, but also many differences from, the British system. One of the main instruments of control over land use and development in France, Germany, The Netherlands and Denmark is a building permit issued by the local authority, and this covers both planning permission and the equivalent of the British building control. Another difference is that in these countries the local plan is a legally binding document. Clear criteria are laid down for the approval or refusal of development permission, and in general, if a proposal accords with the local plan and building regulations, it cannot be refused. This provides relatively firm guidelines and a degree of certainty for both developers and the existing users of land. Many of the rubrics are very detailed: for example, in France the *code d'urbanisme* of 1954, provides a lengthy consolidated list of legal powers and planning rules, and the local plan (*plan d'occupation des sols*) lays down zones of different land use together with other restrictions such as plot ratios. A comprehensive comparison of many West European planning systems has recently been published (Department of the Environment 1989) and Table 2.1 summarises some of the key development control provisions.

In the USA the prevailing values in land use are those of private initiative and free market forces, but an increasing package of federal, state and local government measures is now involved. Garrett (1987) argued that legal considerations are important alongside economic ones in any analysis of urban land use in North America. Although the free market had, and still has, a strong hand in shaping American cities, it is possible to see, from the earliest

Table 2.1 European development control plans

	England	Denmark	France	West Germany	Netherlands
Name of plan	Local plan	lokalplan	plan d'occupation des sols	Bebauugsplan	Bestemmingplan
Coverage	About 25% of the country	Widespread	80% of urban communes	Widespread	Virtually complete
Function	Guidelines for development and land use co-ordination	Implementation of structure plans and development control	Consolidated statement of planning restrictions	Binding land use plan	Statement of planned development and control
Type	Base maps with proposals for specific development and restrictions. Written statement	Detailed land use plan and regulations	Zoning map with land use and regulations for each zone	Land use plan permitted uses densities	Land use map detailed regulations
Basic instrument	Planning permission	Building permit (Byggetilladelse)	Building permit (Permis de construire)	Building, permit (Baugenehmigung)	Building permit (Bouwvergunning)
Decision by	District council planning committee	Technical and environmental committee	Mayor	Chief Planning officer in Kreis/Kreisfreie Stadt	Municipal Executive

Source: DoE 1989

days of large scale urban development, that a degree of centralised planning or regulation has been important. This can be seen especially in the widespread use of the grid plan which imposed an ordered pattern upon subsequent land use and urban morphology. Pierre l'Enfant's plan for Washington in 1791 for example, and the New York grid plan of 1881 have both left indelible marks on these two great cities. Concerns for health, order, clear street lines, straightforward land transactions and recognisable property lines were all important considerations in the original choice of the simple grid plan.

One of the key features of the relatively weak present day land use planning in the USA is the widespread use of zoning, a notion which partly reflects a distrust of the powers of a completely free market to allocate land effectively (Garrett 1987). Zoning of land use in the cities of the USA stems from the period between 1916 and 1926 when it was first instituted to relieve congestion of traffic and land uses within the garment district of New York. Following an important test case at Euclid, Ohio, in 1922, zoning ordinances spread rapidly, being adopted in some form within all forty-eight states by 1946 (Goldberg and Chinloy 1984) and by 9,000 local governments in 1968 (Garrett 1987). At that stage, every major city, except Houston had accepted zoning.

The basic intention of zoning is to regulate land use and intensity, but in many cases it goes much further and includes layouts, plot sizes and subdivision, setbacks, shadows, aesthetics and housing type. Some idea of the scope can be gained from the very condensed summary of the zoning ordinances used by the city of Philadelphia in the 1970s which is shown in Table 2.2. Despite the technical appearances of such regulations it can be argued that their main purpose is to express the taken for granted understanding of social order (Perin 1977), the subtext is homogeneity associated with the lowering of social conflict and the prevention of negative spillover effects. This argument then is rather different from the traditional explanations of land use planning in the UK, where the protection of weaker groups in the community and the promotion of social equity are important goals. In the UK a form of zoning is applied to land use patterns in local plans, whereby some attempt is made to keep non-conforming uses separate. Thus industry will normally be removed from, or prevented from developing in areas designated for residential development. In the USA it seems that land use zoning is at least largely about protecting individual property rights and reducing investment uncertainty by transferring some of the risk to the community. An example of this is cited by Fischel (1985) who claims that the antipathy of wealthy suburbanites to low income housing is not based upon aesthetics or the physical nature of the land use, but upon social status and especially a fear of crime.

Zoning is not the only regulatory tool affecting land use in the USA. Since 1970 there has been an extension of centralised land use regulation at state, regional and federal level and although this has a limited effect upon cities, it

Table 2.2 Philadelphia zoning ordinance: summary

Scope: The district requirement consists of 50 different classifications, 28 residential, 10 commercial, 8 industrial and 4 special districts.

Permitted uses: Districts are ranked from most to least restrictive, and generally the uses permitted in the most restrictive district are also allowed in less restricted areas. In the residential category permitted uses range from single family detached dwellings, through duplex and multiple family dwellings, to group housing, multiple housing, conversion districts and finally a limited mix of residential and commercial activities.

Yards: These are open, unoccupied areas between buildings and property lines. Regulations may stipulate size and setback in all kinds of district in order to improve safety and visibility.

Lot area and width: Regulations to prevent construction of buildings in lots which are 'too small for reasonable use'. Examples of minima are plots of 10,000 square feet in residential category R-1, and 1,440 square feet for multiple family dwellings in R-10.

Occupied and open area: A limitation on the proportion of the lot which may be built upon, to prevent overcrowding and to provide amenity space. E.g. 35% of the plot in R-1, 70% in R-10.

Maximum height: Restrictions on the erection of high buildings which rob neighbours of light or air, or which cause congestion on public streets.

Gross floor area: A regulation designed to control building height and bulk. It is a ratio based upon floor area, plot size and height. In the CBD the ratio may (with bonuses for setbacks and open space) go up to 1,200%.

Density controls: Zoning ordinances often specify maximum residential density in terms of persons per acre. Philadelphia does this indirectly through the above measures.

Source: A Guide to the Philadelphia Zoning Ordinance, Citizens Council on City Planning, 1970, Philadelphia

has grown steadily and it does show political staying power (Popper 1988). More important has been the parallel emergence of new regulatory devices at local level. In the late 1960s, land reform processes persuaded most local communities to adopt a comprehensive development or growth management plan, and although these have no statutory power they are increasingly being used for land use regulation at local level.

In the USA, as in Britain, some of the most severe challenges to planning have come in the run-down inner areas of the largest, decaying industrial cities. It is here that the land market has been in most obvious difficulties, although whether this is due to intrinsic market weaknesses or a gross distortion of the market by bureaucratic and other public intervention is debatable. At the very least it seems that the private sector has been deterred by the size and long time scale of the problem. The responses in both countries, with Britain often copying American policy, have had many similarities. Land use planning has been given a major role in the processes of job creation and revitalising the urban economy and environment. New development bodies have emerged within the context of a public–private partnership. In the USA, the role of such partnerships in rehabilitating districts of Baltimore and Pittsburgh, is well known, and the principles have been applied more recently to long-standing problem areas in a number of British cities including dockland districts in London, Manchester and Liverpool. Land use planning is only one part of a much larger scheme of revitalisation in cases such as these, but the important part which it plays has recently been reviewed by Lawless and Ramsden (1990) in the city of Sheffield.

In summary, it is possible to see that planning has a wide range of instruments for land use regulation, especially in European cities. Its operation is as much concerned with politics and social values as with economics, and for this reason it has been challenged from many directions, particularly during the 1980s when the tide seemed so often to be running against public intervention. During that decade, and currently, much of the emphasis of planning has been changed, away from regulation towards the promotion of development. Much has been expected of it, from the resolution of small scale conflicts over competing land uses to a key role in the regeneration of flagging urban economies. Certainly, there are isolated successes to be recorded, but overall it must be seen as a relatively small piece in the overall jigsaw of urban land use. The theme of the public regulation of land will be returned to in the broader context of land use policy in Chapter 6.

Fiscal measures

The state may intervene in the urban land market through many fiscal measures. Essentially there are three kinds: (1) routine raising of revenue, (2) taxes or levies on land and property in order to recoup some of the enhanced value of the developed land which is considered to have been created by the community, and (3) subsidies to promote development or encourage important activities.

Routine revenue raising by means of property taxes takes a number of forms in Britain. Up until 1990, rates were levied on all property as a basic method of financing local authorities. Although they were replaced in 1989 by the Community Charge for the residential sector, a national non-domestic rate based upon notional rental values is still levied on industrial and commercial premises. Notice has been given that this Community Charge will, in turn, shortly be replaced by a new Council Charge, to be based upon domestic property values. Elsewhere, in parts of the USA for example, capital values of property are used as the basis of local taxes, and in New Zealand and Denmark, site value rating is used. The precise operation of the local land and property tax systems can have a marked effect upon attitudes towards development. For example in Britain, the absence of a tax on land which is vacant lessens the pressure to develop it, in France there is a tax regime which discourages preparing land for development, and in Japan the tax system makes it advantageous to hold financial assets in the form of land (Wijers 1988).

Major debate revolves around the extent to which the profit accruing from land development should be taxed. On the one hand is the view that the developer should be allowed to keep any profit gained from developing land to a higher use. On the other hand is the argument that it is the community which creates enhanced land values, through the general expansion of the city, or through specific actions such as granting planning permission or the provision of infrastructure, and therefore it is the community which should reap the benefit. Whilst it is true that land owners can improve the value of their land by their own actions, the value may also rise even though they do nothing. A middle ground exists by taxing the betterment in order to return some of the financial gain to the community. Basically this system, operating through changing grades of betterment levies, capital gains taxes and development land taxes, has been used in the UK since 1947. Depending on the level of taxation there are benefits for all, since the state's role is financed and the land owners keep some of the betterment, thus providing incentive for the market to continue.

Subsidies to encourage desired forms of development are the other part of the fiscal measures, and these too take a number of forms. For example, subsidies may be applied to transport to encourage the efficient operation of the urban economy, to such features as street lighting where it is administratively difficult to collect a charge, to sewerage and water supply which are capital intensive and where there may be good reasons for a state monopoly, and to education and health in order to encourage a high level of consumption. Few of these subsidies are aimed at urban land per se but they all have important implications for it. In Britain, one of the most substantial and relevant forms of subsidy is that available on housing, with income tax relief on mortgage interest for privately purchased housing and direct building and rent subsidies in the public sector. More specifically, subsidies are available in areas of development difficulty in many European countries, and to a lesser extent in the USA, through such devices as income tax and rating relief or grants to encourage investment and job creation.

CONCLUSION

In reviewing the mechanisms by which land use is allocated in the contemporary western city, we return to the original dichotomy between market forces and public intervention. And yet it is not a wholly realistic dichotomy, for in cities of all the western nations it is not a case of either one or the other, but a question of the balance between them. There is an informal spectrum running from European countries such as Britain, France and The Netherlands, where planning is well developed, to the USA where it is fragmented, relatively poorly developed and has only weak control over the market. It is also true that in the 1980s political rhetoric downgraded public intervention, but in many cases the reality did not bear this out. In some cases public intervention could be seen to be used to control the market, but in other cases, even in the same country, it was used to stimulate the market. We must be careful not to carry the distinction too far, for as Blowers (1986) pointed out, town planning is essentially a mode of decision making for the allocation of land uses in a system where the market is the primary mode.

The operation of that market in urban land also needs to be qualified. It tends not to operate quite according to the economist's view of normal markets. This is because land is an unusual commodity and because there are large elements of tradition, sentiment and other non-pecuniary factors involved in its sale. The market is also constrained by the need for the public to provide much of the infrastructure and by the fact that developers necessarily operate according to relatively short time scales whereas the community takes a longer term view.

The pattern of land use we see in today's cities is the cumulative result of many generations of development and there are large elements of inertia. Even so, the nature and form of the city has changed rapidly, especially in the past quarter of a century. It is these changes which show up some of the shortcomings and oversimplifications of the economists' models of urban land use, especially those based upon the highly restrictive assumptions of bid-rent theory. Even the best of these models can only explain some of the land use patterns and processes in some of the cities for some of the time. Above all, it is the assumed role of the city centre and the reliance of these models upon its dominance which is fundamentally flawed.

The CBD still exists, but as people, job, shops and social life have decentralised, the city centre has become less important as the organising focus for

41

the city and the urban land value gradient has consequently changed profoundly. The motor car has restructured the city, not simply by adding more suburbs around the centre, but by producing a complex outer city with its own focal points and a large measure of self-sufficiency. It offers a wide range of urban activities without singular urban concentration. As the impact of the motor car upon the city has become consolidated, and now, in turn, increasingly questioned, a new set of influences is beginning to produce a new wave of changes. Some of these changes will be examined in later chapters, but first it is necessary to pause and investigate how fully and precisely we can actually measure patterns of urban land use. This will be the purpose of the following chapter.

3

MEASURING AND
MONITORING URBAN LAND

INTRODUCTION

The need for careful land use planning to be based upon a firm knowledge of existing land use patterns is indisputable, especially in such small and highly urbanised countries as Britain and some of its European neighbours. Despite this, the state of our knowledge of urban land use is still far from satisfactory. Even the most basic facts are often contentious. Numerous independent commentators have drawn attention to the shortage and patchiness of urban land use data, and the way in which this handicaps effective land planning and allocation, yet paradoxically the acquisition of land use statistics does not appear to be a top priority for local authorities, at least in Britain. Coppock's judgement that 'the collection of adequate data on urban land use and land use changes is always likely to present difficulties' (Coppock 1978: 55) unfortunately remains true.

Although recent years have not seen great improvements in the availability of information, they have seen major changes in both the planning and technical contexts of land use studies. Planning in the UK evolved from its domination by development plans and highly detailed land use maps in the 1940s and 1950s to a more generalised system of structure plans in a new local government framework after 1974. By the mid 1980s even the validity of these structure plans was being questioned and a new system of unitary plans was proposed at the end of the decade. Greater attention was also given to a number of specific land use issues, notably those in the inner city and the greenbelt and those concerning vacant land and land for houses. During the same period the technical context also changed as traditional ground surveys were supplemented by aerial photography and remote sensing, and as manual analysis and draughting gave way to digital mapping, computerised land management systems and geographical information systems.

Even with these advances, the stark fact remains that the best available comprehensive survey of urban land use in England and Wales is based upon 1969 aerial photography (DoE 1978). It contains only five crude categories (mainly residential, mainly industrial and commercial, educational

43

and/or community, transport and open space) and measures only areas of developed land of 5 ha and above.

PROCEDURAL MATTERS

The collection and analysis of urban land use data presents immense problems, not least because as Coleman (1980) observed, land use survey in the UK is not entrusted to a unified professional organisation. Instead, it is split among a multiplicity of planning authorities with only a modest degree of central control being provided by the Department of the Environment. In many cases the purpose of survey, and the definitions and classifications used vary from one local authority to another and there is no overall consistency of either input or output. In order to understand what is available, and some of the constraints, it is useful to consider a few preliminary matters.

Purpose of survey

Research workers using existing land use data encounter the problem that the nature of the data is heavily influenced by the purpose for which it was originally collected and this inevitably limits its more general utility. Broadly speaking, three uses have governed such exercises. First, a number of studies have been undertaken to measure the overall extent and expansion of urban areas. Such studies are commonly large scale, being national or regional in scope, but rely upon a very coarse subdivision of perhaps no more than half a dozen categories. Second, are the inventory type of land use exercises under-taken mainly by local authorities to help them analyse and monitor their planning policies, or as part of a more general property/land management system, or for related needs such as rating purposes. Third, are the more specific, subject based surveys relating to such problems as derelict/vacant land or the availability of housing land. Within the public sector the main guidance on land use statistics comes from the DoE which spells out its requirements to local authorities through periodic circulars. This process gives some degree of comparability amongst the data, but it remains the general case that land use surveys in the UK have been compiled by different agencies, at different times using different, often incompatible, techniques.

Defining urban

Defining what is meant by urban presents a number of other initial diffi-culties. The emphasis upon local authorities as the primary collectors and users of land use information gives the exercise an immediate framework of administrative areas. This is far from ideal, for administrative boundaries rarely coincide with the physical extent of urban growth and large urban agglomerations are commonly divided between a number of separate authorities. Attempts to refine the definition have included the imposition of a population size threshold (Best 1981; Guerin and Mouillart 1983), consideration of continuous developed areas covered by buildings and urban structures (DoE 1978), the use of a residual urban definition from agricultural surveys (Best and Anderson 1984; Deane 1986) and attempts to generalise urban/rural boundaries by statistical techniques (Ward 1983).

Nature of urban land

The particular nature of urban land poses difficulties at two levels, conceptual and practical. At a conceptual level it is important to consider the overall political economy. It is commonly assumed that a rational pattern of land use evolves, mainly by activities competing for sites through the process of supply and demand, yet it is equally clear that the urban land market functions imperfectly. The balance between public and private sectors has shifted in recent years and many external features such as inflation, credit availability, social change and growing affluence have produced additional turbulence in the patterns of urban land use. Old buildings survive alongside new, and vacant land persists alongside intensively developed sites. The net result is that a difference exists between the observable land use of a given plot and its potential in a planning context.

At a practical level, the main problem is that urban land forms a very dense and small scale mosaic of development. Questions arise over the choice of the basic unit for survey. This may be decided on grounds of cost, in which case some sort of grid overlay or sampling may be appropriate (Dickenson and Shaw 1982) or there may be a technical constraint as in the case of the level of resolution applicable to remote sensing. For a detailed survey however it is desirable to consider individual properties (Coppock 1978); or curtilages including the land attached to buildings (Dickenson and Shaw 1977). Even within individual curtilages there may be several land uses. In such cases it is normally appropriate to record the principal use, i.e. that use upon which all others depend for their existence.

Classification schemes

To allow order or patterns to be recognised, a system of classification is needed. Not surprisingly, no ideal system of land use classifications exists and it is unlikely that one could ever be devised. In practice, most schemes are not classification (*sensu stricto*), where individual observations are grouped on the basis of similarities, but rather a form of discriminant analysis where each observation is compared with an *a priori* scheme, and pigeon holed accordingly. Frequently there is conflict over the number of classes used. A small number of classes gives ease of allocation but much loss of information and a large number of classes becomes confusing and unwieldy. The ideal requirements for classification schemes have been outlined by Rhind and Hudson (1980) and Hill (1984).

A distinction may initially be made between, on the one hand, land form or cover, and on the other land function or activity. Form or cover is essentially the nature of the elements in the landscape: for example, types of buildings, structures or open spaces; whereas function or activity concerns what the land is actually used for. The distinction is important because it relates to the methods of gathering information. For example, land cover may be discernible from remote sensing imagery, but because cover does not give a reliable guide to activity, the latter normally requires a ground survey or documentary evidence.

Most of the large, general purpose, classification schemes provide poorly for urban land uses. The Second Land Utilisation Survey gives only 4 classes (out of 13) to broadly urban uses and more recent schemes designed for use with remote sensing techniques are even less discriminatory. The United States Geological Survey, for example, has only one urban category out of 8 first level groups (Anderson 1976) and the classification proposed for the European CORINE (Co-ordinated Information on the European Environment) project has one category of 'Built up and Related Areas' in a group of 8.

An attempt to standardise the individual classification schemes devised by British local authorities in the 1940s and 1950s was not made until the mid 1970s when the National Land Use Classification was promoted (NLUC 1975). Like most others this scheme is hierarchical. It has 15 major orders, 78 groups and 150 subgroups and some compatibility exists with the Standard Industrial Classification (Markowski 1982). Dickenson and Shaw (1977) considered applying this scheme in their study of Leeds, but concluded that it had a number of significant shortcomings for use in urban areas. Also in the mid 1970s the DoE attempted to collect statistics on land use changes from local authorities, but that exercise was largely unsuccessful (Dickenson and Shaw 1982; Sellwood 1987). A renewed attempt, on a different basis was started in 1984 (DoE 1986) using the classification in Fig 3.1.

In a more detailed sense, and for the application of town planning legislation, all land is deemed to have a use, as defined in the Use Classes Orders of 1972 and 1987 and a comprehensive gazetteer of these has been produced by Godfree (1988).

URBAN LAND USE STATISTICS

Sources

The sources of statistics on urban land use in the UK are many and varied, but unfortunately they do not add up to either a coherent or a comprehensive coverage. Valuable summaries of available sources have been made by

Coppock (1978), Gebbett (1978) and Best (1981), but there have been significant developments since then in the four spheres discussed below.

Maps and ground surveys

The very comprehensive sets of topographic maps produced by the Ordnance Survey are a useful basis for measuring the overall extent of urban areas (Fordham 1974) as well as providing more specific land use information. Changes noted by Ordnance Surveyors as a part of their regular work on map revision form the bases of attempts by the DoE to monitor land use changes (DoE 1986; 1987; 1988a). Despite the accuracy of the OS maps, some doubts must surround the quality of the land use data thus gathered. In particular, there are questions about how systematically the information is gathered and the variability of the time lags between changes taking place and being recorded.

Detailed land use maps compiled from ground survey formed the basis of

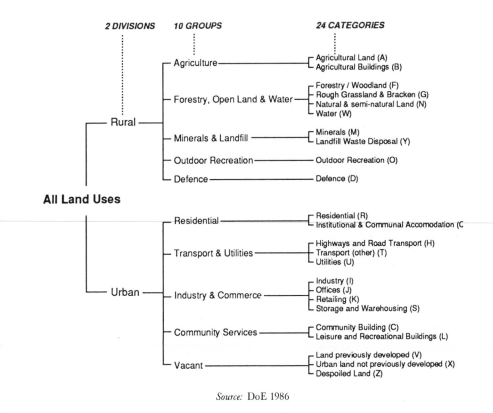

Source: DoE 1986

Figure 3.1 Land use change statistics: classification structure 1986

47

both the first and second land utilisation surveys and of the many town maps which were required by the postwar planning system. The latter were extensively used by Champion (1974), Best (1981) and others to derive their estimates of the extent of urban land, but, as these authors point out, the town maps were subject to many inaccuracies and inconsistencies and by 1970 they ceased to have any contemporary purpose. Land use maps compiled from ground surveys have a number of advantages, they are detailed, accurate and direct records but they are also cumbersome, expensive, difficult to analyse and present only a static picture. For these reasons their use today is mainly restricted to ad hoc surveys, for example, of derelict land, or they are used selectively to check and calibrate other methods of gathering information.

An increasing amount of information is now becoming available in digital form. Little of it is direct land use information, but it has a growing utility as the basis for land use investigations. In Britain, for example, the Ordnance Survey currently provide maps in vector data form for most of the country at scales of 1:2,500 and 1:1,250, and for selected areas at a scale of 1:250,000. Maps in Raster data form are available for England and Wales at a scale of 1:10,000. For large settlements a 1981 Urban Areas boundary set, digitized from 1:50,000 Ordnance Survey maps is held by the Department of the Environment. In the United States, the Bureau of the Census releases computerised map boundaries for the Topologically Integrated Encoding and Referencing System (TIGER). These merge the street map boundaries with shapes, points and identifiable features such as roads, railways and waterways, from the US Geological Survey (Levine 1990).

Aerial photography and remote sensing

The use of aerial photography for land use survey and urban analysis has been well established since the 1940s and Berlin (1971) has provided a useful bibliography of early studies. From the early 1970s greater attention was being paid to the use of sequential photography to monitor urban change (Dueker and Horton 1971; Hathout 1988) especially in North America, but some British local authorities began to use aerial photographs in conjunction with other data such as that from the census. Improvements have been made possible by advances in photography (Lindgren 1985), but the quality of the land use information remains limited by the nature of urban areas and by the amount and kind of data an aerial survey can provide.

Dickenson and Shaw (1977; 1982) used aerial photography extensively in their study of Leeds and panchromatic photography taken in 1969 at a scale of 1:60,000 was used in the DoE's comprehensive study of developed urban land (DoE 1978). The latest air photograph survey of land use in England and Wales was carried out in 1984 (Hunting Surveys 1986), but many local authorities also commissioned cover of their own area in conjunction with the national census in 1991. More specifically, the use of aerial photography in studies of derelict land has been explored by Collins and Bush (1974) and a review of remote sensing in this context will be found in Kivell *et al.* (1989).

New methods of gathering land use information were introduced with the launching of the first LANDSAT satellite in 1972, but the imagery was, and still is, limited by the poor level of spatial resolution of the sensors and the complexity of the urban scene. Significant advances have been made recently, especially with imagery from the French SPOT satellite which is capable of resolution down to 10 metres in the panchromatic mode and 20 metres in multi-spectral mode. (Broadly speaking, panchromatic covers the visible part of the spectrum whereas the multi-spectral mode is any reflected radiation including the visible and infrared bands.) The application of remote sensing to studies of urban land use has recently been reviewed by Harrison and Richards (1987) and Whitehouse (1989) who suggested that, as yet, such techniques offer poor accuracy in built up areas. Conversely, high levels of classification accuracy using SPOT-1 imagery were claimed by Collins and Barnsley (1988), but their classes distinguished only between high density residential, low density residential and commercial/residential uses. More recently, an investigation into the use of LANDSAT and SPOT imagery for measuring land use change has been conducted at the National Remote Sensing Centre in Farnborough. Three techniques for mapping urban change were evaluated; visual interpretation of enhanced hard copy imagery, multi-spectral classification and image differencing. Visual interpretation was found to be the most sensitive, with an accuracy of 98.7 per cent obtainable from SPOT imagery; however, it was a slow and costly technique and the non-digital form of the map outputs limited further analysis or the incorporation of the data into a Geographical Information System (National Remote Sensing Centre 1989).

Remote sensing (including aerial photography) can offer a number of advantages including speed and cost-effectiveness, the ease of time sequence comparisons and the ability to overcome site access problems. Alongside these must be set a number of disadvantages including low levels of resolution, the limited success of land use classification algorithms, the variety of responses given by land under different conditions, the imperfect relationship between land use and land cover and the need for sophisticated equipment to analyse data. Aerial photographs are widely used, but mainly for specific purposes such as individual site evaluation rather than for compiling comprehensive land use inventories. Satellite imagery is currently little used, but as techniques improve its potential will be increasingly realised. A brief list of sources is provided in the Appendix.

Published statistics

A number of government bodies and departments publish statistics on land use, but only a fraction of them relate directly to urban areas. The publication of sequential statistics on agricultural land permits crude residual estimates to be made about the total amount of urban land (Best 1981), and allows the calculation of five-yearly moving averages of transfers from agricultural to urban uses (Sellwood 1987). These changes may also be monitored more directly for the mid 1980s from the DoE figures of land use change referred to above. An indirect view of the composition of urban land use may be gained from statistics collected for rating purposes by the valuation office of the Inland Revenue (Central Statistical Office 1990). These provide a breakdown into seven broad categories, but they refer simply to hereditaments and give no indication of relative sizes or areas. In any case, as explained below, the whole rating system is currently being changed.

A limited amount of information is also published on more specific land use activities. Statistics on derelict land are published sporadically (DoE 1974; 1982; 1991) and a rolling register of publicly owned vacant land is maintained by the DoE (see below). At one time the DoE also used to publish commercial and industrial floorspace statistics but this series ceased in the mid 1980s. In addition, there exist a number of other mainstream data sources on employment and population which may be used with mixed success, as Rhind and Hudson (1980) and Champion (1972) have shown, for indirect measurements of urban land.

All of the above sources have shortcomings in terms of the directness of their relationship to urban land use, their level of aggregation, the partial nature of their coverage or the continuity of their data.

Local authority administrative sources

A number of local authority departments collect a wealth of data which can be used in the study of urban land use. Most obviously, routine planning statistics relating to allocated uses, development control and planning applications yield much information at local level. These, for example, were the basis of the DoE's unsuccessful attempt to gather annual statistics on land use change in the mid 1970s and they are used today, rather more successfully, as important inputs to local authority land information and management systems. Architects and engineers departments collect information on building starts, completions and demolitions, and sundry others such as building inspectors and education departments gather data relating to land and property for their own purposes. Clearly, the effective use of such information requires a high degree of co-operation and integration, but it offers the potential of a high quality, detailed local database which can be updated regularly. Its use for any kind of national inventory is, however, severely limited by wide differences in classification, definitions and the timing of surveys (Champion and Markowski 1985).

Other local authority sources include rating lists, although relatively little use has been made of these (NERC 1978) and substantial changes in their organisation are under way. In Scotland, domestic rates were replaced by the Community Charge in 1988 and in England and Wales the same change took place a year later. Effectively, these changes made the individual person, rather than the individual property, the basis for levying local authority rates. As a result, the scheme produced lists of Community Charge payers which had no value for land use studies. In 1991, the government tacitly accepted the unpopularity and inefficiency of the Community Charge and announced plans to replace it with a new Council Charge which will be levied upon the capital value of property divided into a series of distinct price bands.

In parallel with this, there is a system of National Non Domestic Rates (NNDR), sometimes called the Unified Business Rate, to cover virtually everything except private residential properties. Under the provisions of the Local Government Finance Act (1988), the valuation officer for the charging authority is required to compile and maintain local non-domestic rating lists. These initially date from April 1990 and will be updated every five years. In addition, a central list, to include government buildings and crown properties will be deposited with the DoE. The local non-domestic ratings lists are available for public inspection at the offices of the district authorities. As with the old lists, they do have some value for land use studies. They contain details of the property type or activity (typically in as many as 150 categories), the address, and the rateable value (based, as before, upon a notional rental value).

Analysing and manipulating data

Once land use statistics have been gathered they need to be processed. In many respects recent advances in manipulation and analysis have been far greater than those in gathering data.

Traditionally, land use data have been treated as a nominal scale variable, amenable to simple tabular and map presentation and only the simplest statistical analysis. The manually coloured map, a mainstay of local authority studies until the early 1970s, fell from favour because it was inflexible, slow and costly to prepare and offered limited scope for analysis (NERC 1978). At first, few alternatives were available, but some authorities began using computer based techniques such as SYMAP (synagraphic mapping) and made progress on digitising basic map boundaries.

Quite rapidly there began to develop more sophisticated techniques for the management, analysis and presentation of large, spatially referenced databanks in the form of land information systems and geographical information systems, and there are now many proprietary packages available. It can be claimed that every department in every local authority has a need for geographical information systems and that many other businesses and institutions will find themselves turning increasingly to GIS useage. As with any computer based systems, GIS are subject to very rapid change and development but a very useful and up-to-date review of their applications for land and property matters will be found in Dale (1991).

Geographical information systems, in the broad sense, have tremendous potential in the handling of data on the use, ownership, value and other aspects of land. The technique does not necessarily make the business of data collection any easier, but it enables the maximum benefit to be derived from existing sources. In particular, GIS can act as a database recording current situations and allowing quick and easy access to large amounts of data, they can facilitate the integration of different data sets and the overlay of different spatial patterns. They can be used to search for particular features, to evaluate alternatives and to project the likely impact or consequence of planning decisons. Dunn *et al.* (1991) have provided a particularly good example, using the county of Devon, of the way in which several kinds of data can be integrated in the measurement of rural to urban land use change. Despite the technical power and sophistication of GIS, for the foreseeable future it is likely that their main utility will be as techniques to support and inform decision makers in planning exercises, rather than as a robotic decision maker to overrule human judgement.

Land information systems and inventories began in North America, notably in Canada, in the early 1960s. Here and elsewhere they were at first largely confined to rural applications (Gierman and MacDonald 1982; Jones 1986) but they were also used in the analysis of patterns of land use change on the urban fringe.

Relatively limited planning application information systems have been in use in the UK since the mid 1970s (Grimshaw 1985); sometimes standing alone and sometimes linked to property information systems or attempts to monitor broader patterns of land development and potential (Brown 1985). More recently, local authorities have increasingly realised the potential of land information systems for specifically urban land use studies.

Humphries (1985) provided a succinct review of land and property information systems and he drew an important distinction between LAMIS type approaches, in which spatially defined areas are used with full boundary digitising, and the gazetteer type which depends on a central index based upon postal addresses and/or map references to site centroids. The Tyne and Wear scheme was an important pioneer of the gazetteer type in the mid 1970s (Charlton and Openshaw 1986) but other approaches followed: for example, in London through the CLUSTER (Central London Land Use and Employment Register) consortium (Markowski 1982; Home 1984), Warwickshire (Grimshaw 1988), Manchester (Bourke and Davies 1988), Birmingham (Gault and Davis 1988) and Kingston (Weights 1988). As with any technology there remain problems of comparability and compatibility, not least over conventions for spatial referencing of the data.

Major technical problems remain too in the analysis of land cover and land use data provided by remote sensing techniques. Whitehouse (1989) and Dawson (1991) both suggested that the traditional spectral classification approaches are inappropriate for the high resolution data produced by the new generation of satellites and they explored new approaches based upon texture and context.

Much of the analytical effort to date has been devoted to the recognition and interpretation of existing patterns, but from a planning viewpoint some indication of developing and future patterns is desirable. Sequential data bases, such as the DoE statistics on land use change, allow changes to be monitored retrospectively, but few attempts have been made to predict future patterns. A rare attempt is that by Charlton and Openshaw (1986) who used linear and multiple regression techniques and models borrowed from demography to forecast land use trends, but they conclude that none of these were satisfactory.

Results

A brief examination of the findings of some recent studies will illustrate what is possible using the sources and techniques discussed above. Substantive, although now rather dated, summaries will be found in Coppock (1978), Rhind and Hudson (1980) and Best (1981).

The conclusions of Best and Anderson (1984) are that in 1981 the UK contained 2.05 million ha of urban land (8.5 per cent of the total) and for England and Wales the figures were 1.76 million ha and 11.7 per cent. In the 1960s the loss of farmland to development had averaged 18,750 ha/yr but this fell to 15,000 ha/yr in the first half of the 1970s and to 10,000 ha/yr in the latter part of the decade. Using different definitions the DoE/Office of Population Censuses and Surveys (OPCS) estimated that 89 per cent of the population of England and Wales were living in urban areas which covered 7.7 per cent of the land area and that the total extent of urban land was 3.31 million ha (DoE 1988b).

The most recent figures for the composition of urban land on a national basis remain those from the 1969 aerial survey (DoE 1978) and these are shown in summary in Table 3.1. A breakdown by local authority district is also available. An indication of the dynamic processes of land use change is given by the statistics collected by the DoE (1986; 1987; 1988a; 1989a), and these will be examined in the next chapter. Other figures from the same source show that on average 45 per cent of land developed for residential purposes had been previously developed or was lying vacant in urban areas, a finding which broadly endorses that of Dickenson and Shaw (1982) in Leeds.

Table 3.1 Developed areas 1969, England and Wales

	%
Developed area as % of total	9.8
Predominantly residential	60.8
Predominantly industrial/commercial	17.5
Predominantly education/community	1.0
Transport	7.2
Urban open space	13.4

Source: Department of the Environment 1978

Thematic studies

The development of the urban economy in recent years, and the related planning policies, have resulted in the need for specialised, problem based land use surveys rather than general purpose inventories.

Amongst the most important of these have been surveys of land which is derelict, vacant or otherwise poorly used. By the early 1970s it became clear that the extent of derelict land was growing alarmingly in urban areas and that it was being created increasingly by the collapse of manufacturing, transport and public utilities rather than by the traditional cause which was mining. In order to monitor the problem, and promote reclamation through derelict land grants, the DoE requested local authorities to undertake detailed ground surveys. Information thus gathered was published in three summary volumes (DoE 1974; 1982; 1991). This theme will be examined in detail in Chapter 7.

Closely related to the problem of derelict land, and in some cases over-lapping with it, is vacant land. Getting this land back into active use has been an important part of the government's programme of inner city regeneration and to this end land registers were introduced in 1980. Land owners in the public sector are required to reappraise their vacant or underused land with a view either to making use of it themselves or putting it on the register which would signify its availability for development. The register is held by the DoE in the form of a computer database, together with maps and documentary details of each site. Local councils and other public bodies also hold details of their own vacant land and there is provision for public access to this informa-tion. At 31 March, 1988 a total of 40,000 ha of vacant land was on the register, with 55 per cent of this being in local authority ownership. A number of studies of vacant land have been undertaken, using local authority and DoE source material (Bruton and Gore 1980; Adams et al. 1988) and a compre-hensive literature review has been provided by Cameron et al. (1988).

Given the importance of derelict and vacant land as a planning issue, it is disappointing to record that official figures suffer from many shortcomings

54

(Chisholm and Kivell 1987), the net result of which is markedly to under-estimate the extent of the problem.

Recent faltering moves towards a recovery of urban economies, together with various social changes have highlighted another major land use theme, relating this time to the availability of land for house building. Two issues in particular have been important: the total availability of land together with its regional pattern and the relationship between land on the urban fringe and vacant sites within the city. Residential land surveys, undertaken jointly by planners and builders, were instituted in Manchester in 1979 and were subsequently extended by circular 9/80 to all English local authorities. The issue has given rise to a number of disputes between planners and developers (McKenzie 1983) and although the government has been encouraging pro-development policies (DoE 1988c), by the end of 1985 fewer than half the counties in England had undertaken joint land availability studies (DoE 1989b). The situation regarding industrial land has been the subject of fewer centralised directives, but most urban authorities maintain records of avail-able sites for their own promotional purposes.

Retailing is another significant land use issue and it is worth noting here that it gives rise to particular problems in land use study because of the rapidity of change and the complexity of uses on ground floor and upper levels. The main sources of information are shopping centre surveys, rating lists and trade directories but detailed plans and associated listings for over one thousand centres are available commercially from Chas. E. Goad Ltd.

LAND OWNERSHIP AND VALUES

Land ownership is important to an understanding of land use and develop-ment, not least because of the vexed relationship between the private and public sectors and because the behaviour of land owners, be they profit maximisers or utility maximisers, profoundly affects the urban development pattern. Most European nations had a register of land ownership by the eighteenth century, but not so the UK. In 1925 a land register was eventually established in England to ease conveyancing procedures but even today it covers only two-thirds of all land. The registers in Scotland and Northern Ireland are less comprehensive but at least they have been open to the public for some time. In England and Wales public access to the Land Registry property records was only made possible for the first time in December 1990, after a twenty-year campaign by the Law Commission and others. Upon payment of a fee (currently £12), a check can be made upon the ownership, the nature of tenure and the financial encumbrances of any property that is registered. At present about 13 million properties are registered, with another 9 million not yet recorded. The majority of the latter are properties which have not changed hands since compulsory registration was introduced in 1937. The need for a full cadastral survey, giving details of land ownership and related matters has been noted by numerous researchers (for example: Bruton and Gore 1981; Norton-Taylor 1982; Chisholm and Kivell 1987), yet the situation remains unsatisfactory. Within the public sector information on ownership is slightly more accessible, but even here the statistics are fragmented and have to be gleaned from many disparate sources such as the property information systems and records of local authorities and the piecemeal records of other public bodies. Despite the importance which publicly owned land has in shaping the morphology and planning of major cities, there exists no comprehensive and accurate record of landholdings by such bodies as central government departments, local authorities, nationalised industries and statutory undertakers. Recent attempts by government to make the public sector more efficient and the privatisation of a number of utilities and nationalised industries have revealed a surprising degree of ignorance about the extent and status of their land holdings. A number of studies suggest that in large urban areas the majority of land is in fact in public ownership, with local authorities commonly owning more than half of the total. In Manchester, for example, Kivell and McKay (1988) identified fourteen significant public sector land-owning bodies which between them accounted for approximately 65 per cent of the city's land. The issue of land ownership will be more fully discussed in Chapter 5.

One further facet of the land use question to be considered is that of land values. Here again the familiar pattern occurs, information is sparse and fragmented, especially in comparison to countries such as Austria, Denmark and Sweden where land-value maps are often used for taxation and other fiscal purposes. In the UK sources of detailed information are handicapped by confidentiality and, at best, data can be obtained only in highly aggregated or small scale sample form. Some of the sources have been summarised by Howes (1980) but essentially there are only two. District valuers and valuation officers of the Inland Revenue regularly produce statistics on site values and capital values. Many of these are reported to the DoE (DoE 1988b) which publishes a few of them regularly, for example, housing land sales and prices. Additionally, private valuers, surveyors and property advisers compile statistics on land values and transactions. Some of this information is published by the firms concerned or through professional journals such as the *Estates Gazette* or *Estates Times*, but usually it represents only a sample of land actually sold.

CONCLUSION

In conclusion, it is difficult to demur from the overall findings of Coppock (1978). The availability of urban land use statistics is still unsatisfactory, coverage is inadequate and patchy and there remain large gaps in our knowledge. Many of these gaps will become obvious from a reading of Chapter 4 which attempts to describe and summarise existing patterns of urban land use and some of the more important contemporary changes.

At the local level there exist a number of sources of urban land use information, notably within the records of district and county authorities. These sources however vary in the reliability and regularity of their cover, most of them collect information for purposes other than dedicated land use studies and the classification systems which they use are frequently incompatible. At a national level the DoE has made a number of attempts to collect and analyse land use data but these attempts suffer from many shortcomings, notably their restriction to sample studies and their reliance upon extremely crude classifications. At a supranational level some influence is now beginning to be felt through European Community activity. The most relevant programme, CORINE, has been underway since 1985 with the purpose of providing information on the environment to assist in policy formulation. Information on land use and land cover from remote sensing sources will form part of this programme, but at present this has a lower priority than data relating to topography, soils and biotopes (Briggs and Mounsey 1989).

Clearly, the particular nature of urban land poses enormous problems in terms of its smallness of scale, the complexity of activities and the importance of human factors such as ownership and the planning context. The different means of gathering information all have shortcomings: ground surveys are expensive and cumbersome, remote sensing techniques are unproven and documentary evidence is fragmented and discontinuous both in time and space. Rapid advances have been made in managing, mapping and analysing information especially through GIS techniques, but the nature and provenance of the raw land use data have seen few such improvements. In the light of this, it is perhaps remarkable that urban planning has functioned as well as it has for over forty years.

APPENDIX A

Aerial photography

Within the public sector the main sources of aerial photography in the UK are as follows:

Central Register of Air Photography
Ordnance Survey
Romsey Road
Maybush
Southampton SO9 4DH

Central Register of Air Photographs for Wales
Welsh Office
Cathays Park
Cardiff CF1 3NW

Central Register of Air Photography
Scottish Development Department
New St Andrews House
St James's Centre
Edinburgh EH1 3SZ

Department of the Environment (NI)
Ordnance Survey of Northern Ireland
83 Ladas Drive
Belfast BT6 9FT

In addition the Air Photo Unit at the Department of the Environment, 2 Marsham Street, London SW1, provides a restricted service for government departments and some other public organisations.

Local authorities and government funded bodies such as Research Council Institutions also hold collections of aerial photography for their own use as do a number of universities and polytechnics. Notable amongst the latter are the Universities of Aston, Bristol, Cambridge, Dundee, Durham, Keele, Reading, Sheffield, Swansea and University College London.

In the commercial sector there are a number of large air survey companies, for example:

Clyde Surveys Limited
Clyde House
Reform Road
Maidenhead
Berkshire SL6 8BU

Huntings
Gate Studios
Station Road
Boreham Wood
Hertfordshire WD6 1EJ

BKS Surveys Limited
Ballycairn Road
Coleraine
County Londonderry BT51 5HZ

JAS Photographic
92–4 Church Road
Mitcham
Surrey CR4 3TD

Cartographic Services
(Southampton) Limited
Landford Manor
Landford
Salisbury
Wiltshire SP5 2EW

Geosurvey International Ltd
Geosurvey House
Orchard Lane
East Molesey
Surrey KT8 0BT

Committee for Aerial Photography
University of Cambridge
Mond Building
Free School Lane
Cambridge CB2 3RF

Many small companies offering smaller format and oblique photography also exist and these may be found in the *Yellow Pages* under 'Aerial photography'.

Satellite imagery

The main sources are:

UK National Point of Contact
Space Department
Q 134 Building
Royal Aircraft Establishment
Farnborough
Hants GU14 6TD

SPOT Image
18 Avenue Edouard-Belin
F31055
Toulouse
France

4

PATTERNS AND CHANGES OF LAND USE

Previous chapters have revealed that there are very few cases where statistics on urban land use, and land use change, have been collected and analysed systemically. There are however a number of patchy and disparate data sources and it is desirable to attempt some collation of these in order to assess what we actually do know about the pattern of urban land use. This chapter will attempt three things: to build up a picture of urban land uses, changes and needs at various national and local levels, to provide a number of case studies which will exemplify some of the main contemporary changes in urban land use, and finally, to offer some tentative explanations for the observed changes.

THE OVERALL EXTENT OF URBAN LAND

Even the most basic attempt to measure what proportion of a country's land is urban encounters enormous problems of definitions and units of measurement. It is no wonder therefore that in the only major studies on a worldwide scale, the UNO Global Reviews of Human Settlements (UNO 1976; 1987), the tables contain many blanks, and the information which is given is so dated and hedged around with qualifications that it is of severely limited value. Subject to these shortcomings, Table 4.1 indicates the proportion of land used for urban purposes in a number of different countries. It confirms that the United Kingdom and West Germany are relatively highly urbanised in terms of land use, and that the United States of America and Japan have relatively low levels. Clearly there are great differences in the sizes of these countries, in the length of their urban histories and in the consequences of land being taken for urban uses. In Japan, for example, due the mountainous nature of much of the country and the need to protect the scarce but fertile land near the coast, the pattern of urban land use is far denser than is normal elsewhere.

Comparisons of urban densities reveal that North American cities are far more extravagant in their use of land than their European counterparts. A sample of 17 West European cities with populations exceeding 1 million showed average densities of 45 persons per hectare (UNO 1976), 14 similar cities in Eastern Europe and the USSR averaged 43 pph, but in the USA, 25 such cities averaged just 17 pph. Selected comparisons for the early 1970s are shown in Table 4.2. In crude terms Japanese cities appear to be most densely populated whilst those of Australia and North America are least dense. Sample evidence from the USA produces a range of values for the relationship between population and land converted to urban uses. For example, a study of 96 counties in the north east (Dill and Otte 1971) suggested that for each population increase of 1,000, an average of 89 ha were converted to urban uses, whereas in 53 fast growing counties across the country, Zeimitz *et al.* (1976) calculated an average of 70 ha/1,000.

Even within Europe there is considerable variation in urban population densities. When considering the 'city proper' the United Nations study (UNO 1987) found that the highest gross densities were in Paris (20,848 people per square km), Naples (10,342) and Milan (8,747). Birmingham (4,444), London (4,182) and Munich (4,125) were near the middle of the range and the lowest levels were found in Gotenburg (953) and Tampere (245). The economic and social characteristics of diffent urban societies, as well as the national availability of land clearly influence urban densities. But so too does city size. Best (1981) demonstrated a clear relationship which he called the density size rule. According to this, as the population size of a settlement increases, the land provision (in ha/1,000 pop.) declines exponentially. Clark (1967) believed that there was a pivotal provision or density towards which urban space standards will converge as high density areas thin out and low density districts gain population. At a global scale there is some indication that this is happening, for nearly all Third World cities are becoming more densely populated whilst those in the west are nearly all thinning out.

Alternative attempts to estimate the extent of urban land in various countries have produced contrasting results. For example, the figures calculated by Hauser (1982), Table 4.3, are mostly higher than the UNO ones. This table illustrates the complications of different definitions, for example three-quarters of the UK land classified as urban here is actually settlement land,

Table 4.1 Proportion of land occupied by urban areas

Country	Date	% urban
Poland	1965	1.1
United Kingdom	1971	7.4
West Germany	1968	9.7
Japan	1965	1.2
USA	1969	1.0

Source: United Nations Organisation 1976

Table 4.2 Selected urban population densities

Country	Date	Gross density (persons/ha)
USA (5 largest cities)	1970	22
USA (50 largest cities)	1970	15
France (cities > 100,000)	1970–3	22
West Germany (cities > 100,000)	1970–3	26
Italy (cities > 100,000)	1970–3	34
Spain (cities > 100,000)	1970–3	65
UK (all urban)	1970	24
Japan (Nagoya)	1973	64
(Osaka)	1973	138
(Tokyo)	1973	54
Australia (Brisbane)	1966	11
(Melbourne)	1966	19
(Sydney)	1966	19

Source: United Nations Organisation 1976

but for Canada the proportion is only one-third. Even when attention is focused upon a single country, there are discrepancies between different measurements. In the USA, for example, the Bureau of the Census (1988) found that in 1986, 16.2 per cent of the country was covered by metropolitan areas – Table 4.4. Clearly this is a generous interpretation of what constitutes urban land, for, as Jackson (1981) pointed out, metropolitan areas contain much non-urban land. His figures suggested that in fact rather less than 10 per cent of the area embraced by Standard Metropolitan Statistical Areas was actually urban. His overall calculation was that 14.1 million ha of land, that is, 1.5 per cent of the total, could be classified as urban or urban related. In contrast to this, an official study (US Soil Conservation Service 1971) suggested that in 1967 urban and built up land occupied 24.7 million ha, or 2.7 per cent of the total. A later study from the same source (US Soil Conservation Service 1979) raised the figures to 36.4 million ha and 4.1 per cent for 1977.

Bureau of the Census figures for 1982 classified approximately 2.4 per cent of the USA as urban or built up land (Table 4.5) but there was great variation from one state to another. A number of states in the north east, for example Connecticut, Massachusetts, New Jersey and Rhode Island, had proportions of urban land close to the high levels of northern Europe, whereas others, including Montana, Nevada, and Wyoming had levels which barely registered. At the scale of individual cities there is also great variation in the extent and coverage of urban land. This is caused especially by differences in the way in which administrative boundaries are drawn. For example, basing figures upon municipal limits in 1980, the area of New York (780 square km) was very similar to that of Kansas City (819 square km) even though its popu-

Table 4.3 Proportion of urban land in selected countries

Country	% urban	Country	% urban
Belgium	14.6	Luxemburg	6.6
Denmark	9.2	Netherlands	15.0
France	4.9	Sweden	3.3
West Germany	11.8	United Kingdom	8.0
Ireland	1.5	Canada	0.6
Italy	4.2	USA	3.0

Source: Hauser 1982

lation was fifteen times as great. Chicago and Fort Worth had very similar municipal areas but Chicago's population was ten times greater.

In the United Kingdom several studies to measure urban land have been undertaken in the past three decades, but as documented in Chapter 3 they suffer from the use of different definitions, data sources and methodologies. Fortunately, some sensible comparisons can be made, largely thanks to the efforts by Best (1981) and Best and Anderson (1984) to pull these disparate studies together, although it is worth noting that they were not primarily concerned with urban land. Their conclusions were that in 1981 the United Kingdom contained 2.05 million ha of urban land (8.5 per cent of the total land surface) and that for England and Wales the figures were 1.76 million ha and 11.7 per cent. Here too there are significant regional differences, as indicated in Table 4.6. The geographical pattern is very much an expression of the past two centuries of urban growth, being rooted in traditional patterns of industrial location, but as we shall see later, the period represented by the table, i.e. from 1960 onwards, was one of profound transformation away from those traditions.

Using different definitions, the Department of the Environment and Office of Population Censuses and Surveys estimated that 89 per cent of the

Table 4.4 Metropolitan areas in the USA

Metropolitan areas			SMSAs[a]		MSAs[b] CMSAs[c]	
	1950	1960	1970	1980	1980	1986
Number	169	212	143	318	281	281
Population (million)	84.9	112.9	139.5	169.4	172.3	184.7
Population as % US total	56.1	63.0	68.6	74.8	76.1	76.6
Land as % US total	5.9	8.9	10.9	16.0	16.2	16.2

Notes: [a]Standard metropolitan statistical area.
[b]Metropolitan statistical area.
[c]Consolidated metropolitan statistical area.
Source: US Bureau of the Census 1988

Table 4.5 United States urban and built up land cover, by state, 1982

State	Urban and built up area (%)	State	Urban and built up area (%)
Alabama	2.7	Montana	0.2
Arizona	1.0	Nebraska	0.8
Arkansas	0.9	Nevada	0.3
California	3.2	New Hampshire	4.0
Colorado	1.0	New Jersey	23.3
Connecticut	18.8	New Mexico	0.3
Delaware	9.8	New York	5.8
Florida	7.4	North Carolina	4.8
Georgia	4.3	North Dakota	0.4
Hawaii	3.0	Ohio	8.3
Idaho	0.4	Oklahoma	0.9
Illinois	5.1	Oregon	0.8
Indiana	5.1	Pennsylvania	7.1
Iowa	1.7	Rhode Island	18.0
Kansas	1.4	South Carolina	4.2
Kentucky	2.5	South Dakota	4.7
Louisiana	2.7	Tennessee	3.7
Maine	1.0	Texas	2.6
Maryland	11.4	Utah	0.5
Massachusetts	16.7	Vermont	1.6
Michigan	5.2	Virginia	4.7
Minnesota	1.7	Washington	2.3
Mississippi	1.9	West Virginia	2.0
Missouri	2.5	Wisconsin	3.1
		Wyoming	0.2
United States	**2.4**		

Source: U.S. Bureau of the Census 1987

population of England and Wales in 1981 were living in urban areas which covered 7.7 per cent of the land area and that the total extent of that urban land was 3.31 million ha (DoE 1988). The most recent attempt at a comprehensive measurement was undertaken by Hunting Surveys who used air photographs covering a sample of 2.5 per cent of England and Wales for the years 1947, 1969 and 1980. This survey, like most reported here, was not specifically addressed to urban land use but looked at a broad spectrum of landscape and land use. The conclusions, reported by Deane (1986), were that built up land, urban open spaces and transport routes covered 9.1 per cent of England and Wales in 1980, compared with 5.7 per cent in 1947, but the use of a restricted number of sample sites resulted in relatively large errors, especially in urban areas. In West Germany, a 1985 survey revealed that 11 per cent of land was devoted to houses, industrial concerns and other built up activities (Volksblatt 1990).

The figures can thus be seen to vary greatly according to the methods of

Table 4.6 Regional distribution of urban land in England and Wales (%)

Region	Source			
	Fordham 1961	Champion 1970	Dept of Env. 1969	Loveless 1985
North West	22	26	22	26
South East	13	19	17	20
West Midlands	12	13	12	13
Yorks/Humberside	10	12	10	12
East Midlands	9	11	9	11
South West	7	8	6	8
East Anglia	4	7	7	7
Northern	7	7	6	7
Wales	3	7	4	7

Source: Best 1981; Loveless 1989

collection and definitions used. Urban activities occupy small, concentrated and very valuable areas of land. Usually however this land is poorly measured and quantified, figures commonly being arrived at as residuals from larger surveys, often with an agricultural bias. What they do allow us to establish, albeit crudely, is that the proportion of land devoted to urban activities in the most highly urbanised major countries, e.g. United Kingdom, West Germany and Netherlands, is in the range from 10–15 per cent. These figures are however unusual and the figure for most of the highly developed nations, e.g. France, Italy, Japan and the USA, remains below 5 per cent.

COMPOSITION OF URBAN LAND

Difficult as it is to arrive at reliable figures for the proportion of land which can be classified as urban, it is even more difficult accurately to subdivide these totals into individual land use categories.

For the USA, Jackson (1981) has provided a crude breakdown of categories for 22 large cities, Table 4.7, which suggests that residential land use is by far the largest category, accounting for approximately two-fifths of the total. This is close to the range of 40–50 percent which is commonly quoted for European cities (Best 1981; Lecoin 1988). The figure for industrial land, at 10 per cent, is the same as that suggested for large American cities by the United Nations Global Review (1976: 80). A slightly more detailed, albeit rather dated, breakdown is shown in Table 4.8 which is taken from Clawson's extensive work on land use in the USA. The table usefully distinguishes between the whole city, as defined by administrative limits, and the developed part.

For England and Wales, it is again the work of Best (1981) which gives the

Table 4.7 Land use in twenty-two North American cities

Use	Percentage of developed land
Residential	39.8
Industrial	10.4
Commercial	5.0
Roads	25.4
Other public uses	19.3

Source: Jackson 1981

most reliable guide to the composition of urban land. Drawing upon the work of Jones (1974), which used data from local authority development plans, Best was able to summarise the composition of urban land for 1961, and that summary appears in Table 4.9. The category of small towns, with restricted commercial and industrial activities, is overwhelmingly dominated by residential land, but for the larger settlements housing is again within the range of 40–50 per cent.

The most recent comprehensive figures for the composition of urban land in the UK remain those from the 1969 aerial survey (DoE 1978), but because of the crude categories adopted and different methods of collecting the information, they are not directly comparable with those discussed above. Table 4.10 is taken from this source and it shows the composition of urban land use in 1969, in five crude categories, for all cities of over 200,000 people. Considerable variation can be seen in the proportion of the administrative area which is actually built upon. In the cases of the relatively generously defined metropolitan counties this ranges from over 86 per cent for Greater London,

Table 4.8 Proportion of urban land in various uses, USA

Land use	Cities of 100,000 and over		Cities of 250,000 and over	
	Whole city	Developed part	Whole city	Developed part
Undeveloped, private	22.3	0	12.5	0
Public, streets	17.5	23.6	18.3	20.9
Private, residential	31.6	40.7	32.3	36.9
commercial	4.1	5.3	4.4	5.0
industrial	4.7	6.1	5.4	6.2
rail	1.7	2.2	2.4	2.7
Public, recreation	4.9	6.3	5.3	6.1
other	8.8	11.4	10.9	12.6

Source: Clawson 1972

Table 4.9 Composition of the urban area in England and Wales, 1961 (%)

Category	All urban land	Large and medium towns	Small towns (< 10,000)	New towns
Housing	49	46	79	50
Industry	5	8	4	9
Open space	12	18	11	19
Education	3	5	2	9
Residual (incl. transport)	31	23	4	13

Source: Best 1981

to less than a quarter for South Yorkshire. For the more narrowly defined city areas, it is Liverpool which tops the list at 92.4 per cent, but in Sheffield and Leeds less than half of the administrative area is developed. The composition of major land use categories shows rather less variation, and the great majority of these cities have between 55 and 65 per cent of their developed area given over to predominantly residential uses. Urban open space typically accounts for between 15 and 25 per cent and industrial/commercial activities for between 10 and 20 per cent.

By comparison with Britain and North America, Japan has a relatively advanced system of land use data collection. In 1974 the National Land Agency was established and from 1975–80 a quarter of the country was covered by land use maps at a scale of 1:25,000. A certain amount of land information has been available in digital form since 1976, there is an annual survey of land use trends and major municipalities produce large scale land use maps every five years (Himiyama and Jitsu 1988). Much of this information is produced ostensibly to serve planning needs, but it also reflects a concern for land values which virtually amounts to a national obsession in Japan. A summary of land use in sixteen major Japanese cities, including Tokyo and Nagoya, suggested that altogether approximately 80 per cent of land within the administratively defined city areas was used for urban activities (Himiyama 1985). A summary of the different uses is given in Table 4.11. As with all the tables in this chapter, comparisons between countries are limited by different definitions, and this is particularly problematical with Japanese data.

Summarising these disparate data as far as possible does allow a few tentative conclusions. It is residential land use which is by far the most common component in all of the cases examined. Typically, it accounts for around 45 to 50 per cent of developed land in British, other European and Japanese cities, but rather less than 40 per cent in North America. Open space is a problematical category because it variously includes public and private parks, institutional grounds, some measure of cultivated land and sometimes

Table 4.10 Composition of developed areas in England and Wales, 1969 (cities exceeding 200,000 population)

Urban area	Total developed area as % of administrative area	A	B	C	D	E
				As % of developed area[a]		
Greater London	86.1	58.8	12.2	0.8	4.0	24.2
Inner London[b]		59.2	17.0	0.9	7.2	15.7
West Midlands	74.3	59.7	18.6	1.0	2.6	18.1
Birmingham	89.9	63.1	13.9	1.5	1.9	19.7
Greater Manchester	49.6	57.3	20.4	0.8	2.7	18.7
Manchester	91.2	57.9	14.8	1.8	4.6	20.9
West Yorkshire	30.1	54.0	19.3	1.1	2.7	22.9
Leeds	41.8	50.3	17.3	1.3	3.3	27.8
Merseyside	54.8	61.4	14.4	1.0	5.9	17.2
Liverpool	92.4	56.9	12.9	1.8	9.4	19.0
South Yorkshire	23.1	51.6	24.4	2.0	4.1	18.0
Sheffield	36.4	54.5	15.9	2.3	2.2	25.1
Tyne and Wear	49.1	59.2	17.9	0.8	6.7	15.4
Newcastle-upon-Tyne	58.0	59.0	12.0	1.6	6.7	20.8
Bristol	80.1	63.4	13.7	1.5	6.5	15.0
Nottingham	88.3	55.0	18.7	2.4	3.0	21.0
Cardiff	52.5	54.3	16.7	2.7	8.3	18.0
Hull	79.3	57.5	11.0	0.7	13.1	17.7
Leicester	77.9	58.7	13.5	1.0	1.6	25.2
Stoke on Trent	78.8	51.4	27.1	0.3	2.3	18.9
Plymouth	63.9	65.7	10.1	1.2	9.2	13.8
Derby	70.1	59.8	20.6	1.4	2.0	16.3
Southampton	89.4	64.3	9.4	0.8	9.3	16.3

Notes:
A Predominantly residential.
B Predominantly industrial/commercial.
C Predominantly education/community/health.
D Transport.
E Urban open space.
[a]Inner London: City of London, Camden, Hackney, Hammersmith, Haringey, Islington, Kensington and Chelsea, Lambeth, Lewisham, Newham, Southwark, Tower Hamlets, Wandsworth, Westminster.
[b]'Developed areas' consist of all areas of continuous development, i.e. covered by bricks and mortar or other structures, including transport features and open spaces primarily for urban use.
Source: Department of the Environment 1978

Table 4.11 Summary of land use in sixteen large Japanese cities, 1975–80

Land use	%
Residential	37.7
Transport	18.2
Industry	4.9
Commerce	4.3
Education	4.1
Vacant	3.3
Offices	0.6
Other urban	26.9
Total urban	78.8
Non-urban	21.2

Source: Hamiyama 1985

vacant sites too. Where it is recorded, open space typically accounts for about one-fifth of the urban area. The average for transport too is around one-fifth, although it is markedly higher in American cities than in Europe. Industrial activities, upon which many of the urban economies of the western world are founded, account for a remarkably small proportion of the total, typically making up between 5 and 10 per cent.

LAND USE CHANGE

As urban areas grow and evolve, naturally many land use changes take place. From time to time the growth of urban land in general and the rate at which it is being converted from agricultural uses gives rise to concern. Claims have been made that the dangers of losing this land, at a national level, have been exaggerated (Best 1981; International Science Review 1982), but concern often remains strong at local level. What has changed, at least in Europe, is the focus of this concern; worries over maintaining food supplies now carry less weight in an era of European Community food surpluses, but they have been replaced by more broadly based concerns about limiting urban growth in order to maintain environmental and social qualities of life.

In the United States there is more land and there have never been significant food shortages, but even so concern is now being expressed. Between 1967 and 1975 there was a huge and accelerating loss of farmland, resulting in 354,000 ha of actual or potential cropland being converted to non-agricultural uses (Volkman 1987). Sample studies showed that 37 per cent of this land was of soil capability classes I and II compared with 17 per cent of all land in these categories. Other, somewhat patchy, evidence from the USA suggests that the rate of conversion of land to urban uses was increasing in the 1970s, when it was slowing in Europe. During a particularly active period of urban expansion, between 1967 and 1977, the total amount of urban land

70

Table 4.12 Conversion of agricultural land to built up land, 1960–80

Country	% of agricultural land	
	1960–70	1970–80
Canada	0.3	0.1
USA	0.8	2.8
Japan	7.3	5.7
New Zealand	0.5	–
Austria	1.8	3.6
Denmark	3.0	1.5
Finland	2.8	0.4
France	1.8	1.1
West Germany	2.5	2.4
Italy	–	2.5
Netherlands	4.3	3.6
Norway	1.5	1.0
Sweden	1.0	1.0
United Kingdom	1.8	0.6

Note: This table is based upon rough estimates and thus only indicates the order of magnitude of the conversion process
Source: OECD 1985

increased by 50 per cent, representing an annual growth of 1.2 million ha (US Soil Conservation Service 1979). Although mainstream town planning is less influential in preventing urban sprawl in the USA than it is in Europe, a number of states are increasingly following the example of California in establishing agricultural preservation districts to limit intensive development.

In Europe it was the 1960s which witnessed widespread demographic and economic growth, and hence urban expansion. Between 1961 and 1971 urban areas increased, on average, by the following amounts each year ('000 ha): West Germany 35.9, Italy 28.9, France 25.4, United Kingdom 19.2 (Hauser 1982). However, by 1970, with widespread economic recession and falling population growth, the rate of land conversion slowed markedly in most countries (Table 4.12) and even in the improving economic climate of the 1980s it appears to have remained relatively low. In Britain it is again Best (1981) who has provided the clearest measures of change in the growth of urban land. He showed that the peak period for transfers from farmland was in the 1930s when it was running at 25,000 ha/yr. From a wartime dip, it recovered to around 15,750 ha/yr in the 1960s, but then fell slightly in the 1970s to 13,000 ha/yr. In the 1980s a further fall took the average down to little more than 7,500 ha/yr. A similar trend, albeit from a higher level, can be observed in Japan where the National Land Agency estimates that the areas converted from forestry and agriculture to urban use averaged 69,900 ha/yr during 1970–72, 47,000 ha in 1975 and 30–34,000 ha/yr between 1981 and 1985.

Table 4.13 Changes in land use, England (recorded in 1988: figures in ha, rounded to nearest 5 ha)

| | | | | | New use | | | |
| | | | | | | | Urban subdivisions | |
Previous use	*Rural total*	*Urban total*	*Resid.*	*Transp.*	*Ind./Comm.*	*Community*	*Vacant*	*All uses*
Rural	10,120	8,315	4,370	1,890	1,005	520	530	18,435
Urban	1,130	8,125	3,740	885	1,940	455	1,110	9,255
All uses	11,250	16,440	8,105	2,775	2,945	970	1,645	27,690
Net change	−7,190	7,190	6,255	1,355	850	400	−1,675	

Note: The information relates only to map changes recorded by Ordnance Surveyors in 1988. Many of these changes will have occurred prior to 1988 and some, particularly those involving rural uses, might have occurred many years earlier. Only a small proportion of land use changes occurring in 1988 will have been recorded.

Source: Department of the Environment 1989

Another source of information on land use change in England and Wales is provided by the statistics collected annually since 1985 in a trial project run by the Department of Environment (DoE 1986; 1987; 1988a; 1989). The statistics are compiled from information gathered by Ordnance Survey field staff in the course of routine map revisions. There are a number of shortcomings, for example it is limited to areas 'where it is economic to survey', and some changes may not be recorded until many months after they have taken place; but, on the other hand, the exercise is undertaken each year and it does provide the only available national indication of the dynamic processes of land use change. Broadly, these show that in England in the mid 1980s, approximately 24,000 ha of land saw a change of use each year. Table 4.13 summarises the situation for 1988. In total, 27,690 ha experienced a change of use but, on only about one-third (34 per cent) were the changes from previously rural land to new urban uses. Rather more than a third (37 per cent) of the changes were from one rural use to another, and rather less than a third (29 per cent) were from one urban use to another. All of the urban categories showed net gains, except for vacant land which declined by almost 1,700 ha. Other figures from the same source showed that on average, 46 per cent of land developed for housing had been previously developed or was lying vacant in urban areas. These figures have been used by Bibby and Shepherd (1990) to study the rates of urbanisation in England. They suggested that the overall area of land likely to change to urban use between 1981 and 2001 is approximately 105,000 ha, representing just 0.8 per cent of England's area. By 2001 it is forecast that around 11 per cent of England is likely to be in urban use.

In conclusion it is fair to summarise that in Britain, most other European countries and Japan, the rate at which land was being converted to urban uses slowed markedly during the 1970s and 1980s, but that in North America it continued to increase. This situation reflects different planning regimes, different national realities concerning land availability and differential performances of the national economies. All of these in turn were reflected in the tendency for house building rates to fall during the period in Europe, whereas they increased in North America (United Nations 1989).

LAND AVAILABILITY

The discussion about the composition of urban land, and the rate at which it is increasing leads naturally to a consideration of the future availability of land to sustain a variety of urban activities. In Britain especially, with its combination of a heavily urbanised society, high living standards and small overall land resources the question of land availability generates vexed debate in many circles. As we have seen above, it is supplies of land for housing and industry which are particularly important; the former because it occupies the largest area in any city and the latter both because it continues to form an important part of most urban economies and because it is currently in something of a state of flux.

In the early 1970s the British government formalised its advice to local authorities regarding sufficient supplies of land for future building through Circulars 10/70 and 102/72. These urged local planners to ensure a supply of land sufficient for five years at then current building rates, and much pioneering work was accomplished in Manchester by ten district councils, the Department of Environment and the Housebuilders Federation (DoE 1979). The exhortations were repeated in Circulars 9/80 and 15/84 whereby local authorities were required to agree statements on land availability jointly with builders and developers. By 1987, joint land availability studies had been completed for thirty-three structure plan areas, but much controversy had been aroused and a number of parties had resorted to the courts to solve their disagreements (Hooper *et al.* 1988). At the heart of the argument is the extent to which the supply of land should be left to the market, or be based upon planned assessments of future requirements.

From the planning point of view, this calculation of future requirements is complicated by a relatively poor understanding of future population growth levels, rates of household formation and the geographical distribution of population, the state of the national economy, social preferences and the myriad other factors which affect the equation. Calculating the supposed need is only one part of the exercise, choosing locations in which it may be satisfied is the other. Here builders and planners have come under increasingly strong pressure to decrease the use of greenfield sites by using derelict or vacant land within urban areas (Coleman 1978). From the house builder's point of view, land availability estimates, even when agreed with planners, are simply statements about land which might, under a variety of assumptions, get built upon. There is a complex pattern of prices, locations, attitudes of owners, costs of development and planning policies which will determine how, and whether, land will move from the available category into the development process. Inner city sites pose particular difficulties in terms of uncertain marketing conditions, the difficulty of assembling land and the higher building costs likely to be incurred (Brisbane 1985).

Nonetheless, in the 1980s private house builders, using reclaimed land, did begin to construct houses for sale in Britain's inner cities for the first time in many decades. In evidence to the House of Commons Environment Committee, the Volume Housebuilders Study Group (representing the ten largest builders), reported that in 1983, 30 per cent of house building took place on inner city and other mainly recycled land within the urban envelope (DoE 1984: 454). In London, where the pressure on land is greatest, 55 per cent of land developed for housing in the period 1976–80 was land already in housing use (DoE 1984: 80).

As will be seen later, even in the absence of population growth, the demand for new homes continues. Even more pointedly, in areas of high localised growth, such as the increasingly urbanised area peripheral to London, the pressure is considerable. In the South East region, the housing stock grew by 13 per cent between 1972 and 1983 and there was evidence that structure plan land allocations were being used up faster than expected (Home 1985).

The supply and availability of industrial land in Britain's cities also gives cause for concern. Unlike the case for residential land there is no agency which collects or collates information, but most local authorities do under- take their own ad hoc surveys. A number of years ago Fothergill and Gudgin (1982) drew attention to the way in which difficulties experienced by indus- trialists trying to expand their floorspace in congested urban areas was a major brake upon Britain's industrial performance. In south London espe- cially, much industry is located in old and congested premises: more than 60 per cent of industrial premises were built before 1939 (London Strategic Policy Unit 1987). Because the average density of industrial land use has declined with new machines and working practices, more floorspace is required just in order to maintain existing employment levels. But industrial land in London and elsewhere is under intense pressure from activities which can pay higher prices. In the mid 1980s, industrial land values in inner London averaged £850,000–£1,000,000/ha, compared with £1,500,000/ha for housing land and much higher prices for offices. Even where land is appar- ently available for industry there are many problems. For example, in 1986, Birmingham, still the urban core of Britain's manufacturing heartland, had 336 ha of industrial land available, but 65 per cent of all sites were below 0.4 ha, only 43 per cent were actually available within twelve months and two- thirds of the area was land which needed extensive redevelopment to bring it back into use (City of Birmingham 1986).

CHANGING URBAN LAND USE: CASE STUDIES

Many of the patterns and trends of land use can be explained in terms of broad economic and social changes. These will be examined in the next section, but first it will be instructive to look at a few case studies which will both give examples of some of the more significant forms of urban land use change, and illustrate some of the causal mechanisms. The overall pattern of land use is, of course, constantly evolving, but it is argued that the past three decades have witnessed particularly large scale changes as new elements have had to be incorporated within the fabric of urban land use, and other elements, long established in particular locations, have rapidly declined.

Docklands

An important, and distinctive element of many cities which grew on the basis of the nineteenth-century wave of industry and commerce, was a dockland district. Even a number of cities well inland, such as Manchester and Duisburg developed large docks by virtue of canal or river navigations. One of the largest dock complexes in the world developed in the east end of London in the nineteenth and early twentieth centuries, but this industrial empire had begun a rapid decline by the mid 1960s. In many ways this decline, and the reasons for it, are symptomatic of much broader processes of urban change in the last quarter of the twentieth century. Similarly, the processes of regeneration have a broader significance, having been to a large extent borrowed from North American experience, but now having spread to many British cities outside of London (see also Chapter 7).

In London, the first major dock closure, the East India in 1967, heralded a period of rapid collapse and prompted a series of abortive plans for renewal of the area. Many causes for the collapse can be cited. Technical changes in shipping and freight transport, outdated working processes, the obsolescence of the docks (many were over 100 years old), an outdated infrastructure and changes in Britain's pattern of international trade all played a part, and again these causes have a wider relevance for other industries and other urban settings. Between 1978 and 1981, a period of stark recession for Britain's large cities, the London docklands lost 27 per cent of their jobs and between 1971 and 1981 the area's population declined by 24 per cent.

In their concerted search for new instruments of urban renewal, the Conservative governments of the 1980s tried many experiments. The establishment of the London Docklands Development Corporation (LDDC) in 1981 created one of the most extensive of these urban renewal schemes. It also gave rise to one of the largest urban redevelopment sites in the world covering a total of 2070 ha (45 per cent of which was derelict) including 180 ha of docks and 90 km of waterfront Figure 4.1.

The LDDC, which was partly modelled on the New Town Development Corporations of a generation earlier, simplified the planning and development process and was able to build upon a number of important catalysts. One of these was the successful experience of waterfront regeneration in a number of American cities, notably Baltimore and Pittsburg (Law 1988). Other more localised catalysts have included the availability of generous government grants for land, building, reclamation and conversion work and the provisions of an Enterprise Zone which covers part of the area. The area has also had a rapidly improving transport and communications infrastructure which includes the Docklands Light Railway, the nascent City Airport and a fibre optic telecommunications ring main, and the fact of a boom in the property and financial sectors which were hemmed in by restrictive locations in the adjacent City of London.

Buoyed up by such advantages, the LDDC area saw the creation by 1990 of 20,000 new jobs (with an eventual target of 150,000), the largest private sector house building programme in the UK with 17,000 houses started or completed and 1.86 million square metres of industrial and commercial

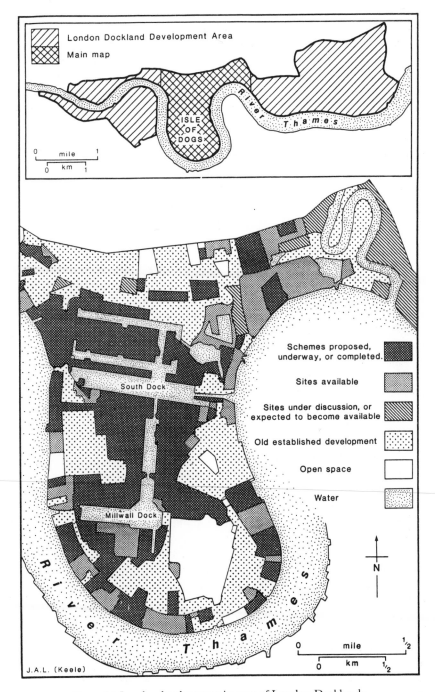

Figure 4.1 Land redevelopment in part of London Docklands

floorspace completed or under development. One development alone, at Canary Wharf, involves 1.1 million square metres of offices, including Britain's tallest skyscraper at 240 metres, and nearly 100,000 square metres of retail floorspace. Because it is located in an Enterprise Zone, this scheme, one of the largest single developments ever in a British city, was subject to only minimal planning control and escaped the need for a public enquiry. In 1992, much of the development remained un-let.

One of the major principles used to encourage development in the dock-lands is that of leverage planning, whereby public sector investment in infra-structure, land assembly and reclamation is designed to attract, or lever, a far greater amount of private sector investment. Measured in these economic terms the device has been largely successful, achieving a leverage ratio in the docklands of 8.7:1 (Brindley *et al.* 1989). This principle was very much in line with the dominant political ideology of the 1980s so, not surprisingly, it was widely used in urban renewal.

The net result of all these changes in land use terms has been immense. A dense mosaic of old housing and congested, dock related, industries and utilities, interspersed with large stretches of derelict land has been very substantially swept away and replaced by a pattern of offices, retail and leisure facilities and both new and converted residential developments. The changes have been far reaching, not only in terms of land use, but also in terms of environmental aesthetics and community structures.

The experiences of London have led to similar, if rather diluted, versions of these renewal plans being applied to docklands in Liverpool/Birkenhead, Manchester/Salford, Teesside, Tyne and Wear, Cardiff, Bristol, Dundee, Hull, Portsmouth, Glasgow and Gloucester, although not all of these have had Development Corporations to guide their renewal.

The success of the LDDC renewal project has been substantial, especially in view of the preceding years of decline and inactivity. Yet the success is by no means absolute. The role of the LDDC has attracted controversy especially where, as a non-elected government agency, it has displaced the elected local authorities. Criticism too has been attracted by the sweeping economic and land use changes, whereby office, retail and leisure activities have replaced, or filled the vacuum created by the loss of jobs and activities based upon the docks, manufacturing and public utilities. The criticisms suggest that the new jobs do not suit the needs of the old community and the new houses are far too expensive for the original residents of the area. Undoubtedly, these arguments will run for some time. Meanwhile the pace of redevelopment in the docklands slackened somewhat as the property market hit one of its periodic slow downs at the end of the 1980s and as it became apparent that there was, in fact, a surplus of office space in London. It is already clear however that the London Docklands development is one of the largest and most profound restructurings of urban land use in the twen-tieth century, a restructuring which in the view of Short (1989) represents not just changes in social and economic relationships, but possibly the beginnings of a new urban order.

Transport modes and airports

One of the most powerful processes prompting changes in urban land use patterns in recent years has been connected with transport developments. Each successive phase of transport technology has brought with it widespread changes in the locational advantages and disadvantages of different cities, as well as more detailed changes in localised urban land uses. In a number of older European cities the remnants of patterns determined by canals in the eighteenth and early nineteenth centuries can still be clearly discerned. Subsequently, the impact of rail transport can be seen to have profoundly affected urban land use, especially through enabling the development of large industrial districts, creating early suburbs and dormitory towns and by facilitating mass access to city centres. Today the railway continues to have an important impact upon urban land use in a number of ways. Some of these are relatively unchanged, but there are two new areas of influence, one connected with decline and the other with growth. The decline involves the release of large areas of surplus railway land in many cities, sometimes, as at King's Cross in London, leading to the possibility of very large scale redevelopment. The growth aspect involves the new generation of high speed trains and links such as the Channel Tunnel which will provide a new economic boost to a few favoured locations.

After rail transport, the next phase of transport technology, and the one which has dominated the twentieth century, was motor transport. It would not be an exaggeration to say that the motor vehicle has been the single most potent force shaping the land use pattern of contemporary cities. In particular it has created suburbs on a scale never before possible, rewritten the rules of urban accessibility, thinned out the overall density of urban development and revolutionised the location of jobs, shops and leisure activities. By its very popularity it has also produced levels of congestion which threaten many aspects of urban life.

In a chronological sense the most recent transport development is that of air travel. Here the influence upon urban land use is not so immediately obvious, but there are two important effects. First, the change from sea to air transport, especially for passengers, has been responsible for both localised and city-wide declines in areas such as Liverpool and Glasgow. Second, the growth of major airports has acted as catalyst for urban growth in general and as a determinant of more localised land use changes.

The influence of airports upon urban areas can be separated into negative and positive effects. Of the former it is noise which is most obvious. A study of Dallas–Fort Worth airport (Young and Schoolmaster 1985) revealed three noise zones with differential effects upon land use. These ranged from minimal effects (zone A) where some sound-proofing was required in sensitive buildings such as schools, through a middle zone (B) where residential building was not advisable but retailing and manufacturing were possible, to zone C where there were severe restrictions on development. On the positive side it was found that population had increased by one-fifth in surrounding counties between 1970 and 1980, and major land use changes had been prompted in the form of large scale commercial, retail and industrial projects, hotels and extensive road development. All local planning officers reported gains to the local economy, with the airport acting as a major stimulus for commercial, especially high-tech development. A similar situation can be observed in Fairfax County, Virginia, where the construction of Dulles Airport thirty years ago has resulted in much of the surrounding area, especially the road from Washington, becoming heavily built up with houses, offices and shopping malls.

In Britain, air transport is less fully developed than it is in North America, but Heathrow, on the edge of London, is one of the largest international airports in the world. Not surprisingly therefore it has had a very powerful effect upon local land use. The airport, which covers 1,200 ha and currently employs a workforce of 53,000, is one of the major focal points in the urban area and has acted as a magnet for manufacturing activities, business parks, hotels and freight depots. In addition to those employed directly at the airport, it is estimated that another 25,000 jobs depend on the airport within a 16 km radius. In 1988–9 Heathrow handled 38 million passengers and cargo to the value of £30 billion (BAA 1990). In turn, all of this economic activity has generated a large scale demand for houses to accommodate the workforce and their families. The extent to which urban growth has been attracted by the airport in recent years can be seen in Figure 4.2. In the same way that a characteristic suite of land uses clustered around the major rail termini of the nineteenth century, so today, major airports act as nodes in the organisation of urban space and attract their own characteristic patterns of activity.

The central business district

As a result of intensely competitive commercial pressures, it is usually in the central area of cities that land and buildings change most frequently. The past three decades in particular have seen major structural changes in the urban core, for example, the decline of old manufacturing industries, technical changes in office and other white collar employment, an intensification of the effects of private transport, and new patterns of retailing. All of these have important land use implications. In fact, there is something of a paradox here because the general pattern of land use in some cases is quite static, for example, prime retailing streets are very resilient, but within the CBD generally the detailed organisation of space and the balance of activities

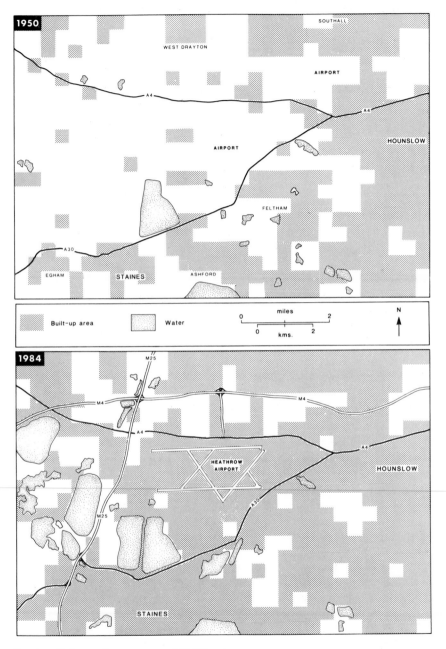

Based on Ordnance Survey maps at 1:25,000 scale.
© Crown copyright

Figure 4.2 Developed land around Heathrow Airport, 1950–84

81

and the ownership of land and property all show almost constant readjustment.

Each city is unique in the way in which it has experienced and responded to these changes, but a number of common patterns are recognisable. Some of these can be illustrated neatly by reference to the city of Stoke-on-Trent in the Midlands of England. Stoke-on-Trent is a city of approximately a quarter of a million people, which forms the core of a polycentric urban area of around half a million. The main shopping and commercial focus is Hanley which is the largest regional shopping centre between Birmingham and Manchester. Most British cities entered the postwar period with congested and old fashioned central areas which were then restructured and modernised over the next forty years. Hanley is unusual in that these major changes were largely concentrated in the period between 1985 and 1990 when internal shortcomings in the layout and facilities of the city centre, together with a need to meet the competition from other retail centres, including out-of-town shopping, prompted a concerted town planning and commercial response. As a consequence, three major elements were added to the centre as Hanley caught up with what had been happening elsewhere. These can be seen clearly in Figure 4.3.

The first major change was the demolition of a number of old buildings on a site covering 2.2 ha, and their replacement by a large, modern covered shopping centre, called the Potteries Shopping Centre, and associated indoor market. This shopping centre provides 31,000 square metres of enclosed retail space, arranged around modern landscaped malls with a naturally lit atrium, a large food court and 1,200 parking spaces. It is a good example of a form of development recognisable throughout large cities in Britain, North America and western Europe. The second addition was a traffic management scheme designed both to ease the internal circulation of vehicles and to speed the flow of through traffic. It consists of a ring road which almost completely encircles the city centre, a number of one-way streets, new car parks and revised public transport arrangements. These changes have permitted the third element to be finalised – the extensive pedestrianisation of certain prime shopping streets, which, together with landscaping and new street furniture, has created a far more attractive and safer environment for shoppers.

These three elements – covered shopping malls, city centre ring roads and pedestrianised streets – exemplified here by the case of Hanley, are representative of similar changes which have taken place in many city centres. The exact nature and timing of them depends upon local circumstances, but the general pattern of the changes, and the commercial and planning processes which have brought them about, have been replicated in city centres across Europe and North America.

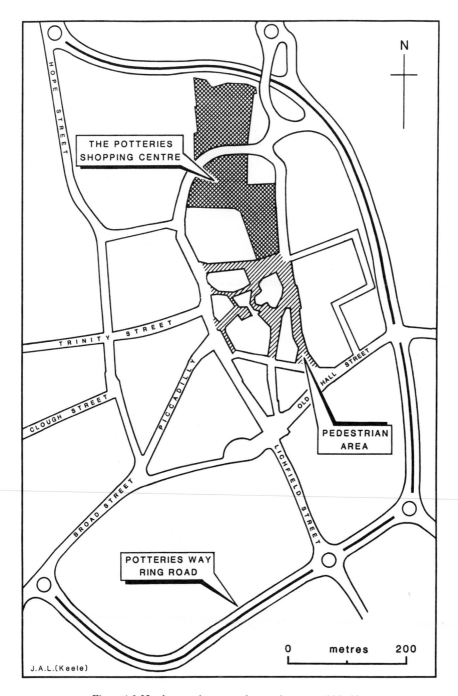

Figure 4.3 Hanley: major central area changes, 1985–90

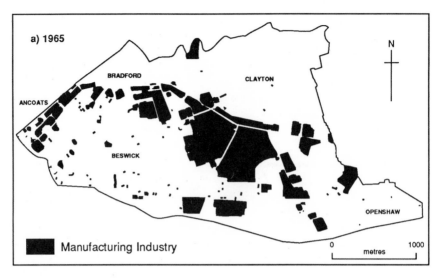

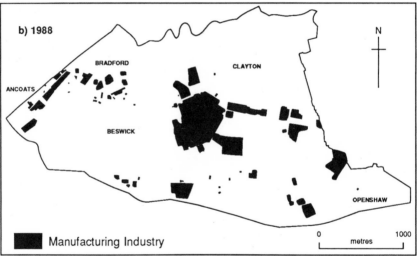

Figure 4.4 Industrial land in East Manchester, 1965 and 1988

Industrial decline

For many decades in western economies there has been a close connection between urban growth and manufacturing industry. It was the Industrial Revolution of the eighteenth century which gave rise to the most substantial wave of urban growth that the world had ever seen, a wave which spread from the United Kingdom and Germany to embrace eventually much of the developed world, indeed, for most purposes, it defined the developed world. After the mid 1960s, however, signs of decline began to show themselves and within a decade many of the largest urban–industrial centres were experiencing a period of savage decline. Between 1960 and 1981 London lost over half of its manufacturing employment, nearly 700,000 jobs, and in the other six conurbations manufacturing jobs fell by 43 per cent (Fothergill *et al.* 1985). This is not the place to examine the reasons for that decline, but it is appropriate to note that one of the most striking consequences was a fall in the number of manufacturing establishments accompanied by important changes in industrial land use. This theme will be returned to in Chapter 7.

Clear examples of the land use consequences of industrial declines can be seen in Manchester, a city with a strong and long manufacturing tradition. A recent study of the Bradford, Beswick and Clayton wards in a heavily industrialised part of east Manchester (Speake 1991), revealed that between 1966 and 1988, the number of manufacturing establishments dropped by 47 per cent. When measured in terms of floorspace, the loss amounted to 57 per cent, over half a million square metres. The extent of change varied from one industrial sector to another, but it was most marked in the traditional industries of the area, notably metals, mechanical and electrical engineering. Most dramatically, in the case of mechanical engineering, the floorspace in 1988 totalled less than one-fifth of that present in 1966. Relatively little floorspace was lost through in situ contraction; the great majority came about through plant closures. In particular, it was the closure of a few large sites which accounted for the greater part of the loss of both jobs and floorspace. The aggregate change between 1966 and 1988 in the pattern of land devoted to manufacturing in the district is shown in Figure 4.4, and this gives an overriding impression of contraction and compaction in industrial land. The original industry was supported by a dense network of public utilities, transport facilities and a large area of terraced housing. Today these other elements have also been thinned, providing a land use pattern composed of remnants of the traditional industries, some small scale new 'opportunistic' activities connected with clothing manufacture and the repair or scrapping of motor cars, a limited range of new housing and more vacant land and public open space than ever before (see Chapter 7).

Two points need to be made to set this brief example into a wider context. First is the point that although this run down of employment and manufacturing floorspace in east Manchester has been particularly severe, it is by no means unique. Similar kinds and scales of decline can be seen in the older industrial quarters of many large cities in western and eastern Europe and the USA. There are some very direct parallels within cities dependent upon the same industries, for example, Sheffield, Pittsburg and the towns of Lorraine in northern France have all experienced similar problems stemming from the decline of their steel and heavy engineering industries. Second, although it is industrial land which is at the centre of the pattern of decline, there is a knock-on effect felt by most other activities. Thus the closure of factories leads not just to the widespread problem of derelict and vacant land (see Chapter 7), but also, indirectly, to an altered pattern of land use in other sectors such as housing and utilities. There are both positive and negative sides to this. For example, there are benefits to be gained from a thinning out of some of the older and denser urban developments, especially where the opportunity can be taken to build modern houses, provide public open space and improve the whole urban environment. However, important though these activites are, they are not wealth producing and no city can exist without activities which provide jobs. In the short run, there is also evidence to suggest that the rate at which industrial land has been falling into disuse exceeds the capacity of most city economies to absorb it into active reuse.

Utilities

In the restructuring and modernisation of the city in the last half of the twentieth century one small, but important, set of activities which has undergone far reaching change is that of public utilities. Some have been made redundant by technical change, such as Britain's local gasworks which have been replaced by a national grid distributing North Sea gas, or the old electricity generating stations which have been overtaken by nuclear power. Equally important are changing economies of scale and the changing pattern of urban transport and accessibility which have rendered many inner city locations less suitable than originally. Wholesale markets for fruit, vegetables and meat, and abattoirs, for example, have been increasingly affected by urban congestion, the shift from rail to road transport and changing retail patterns. Many have closed, or moved out to new locations and a number of the sites which have become available for redevelopment in this way have been in highly visible or sensitive locations. For this reason, their reuse has often led to intense conflict between commercial interests, politicians and local community groups. Intense and protracted planning difficulties have often resulted, for example, at Covent Garden in central London and Les Halles in Paris.

Paris has seen a number of such redevelopments, many of which have led to acrimonious disputes in a city where the quality of the urban environment has been better safeguarded than most. Burtenshaw and Moon (1985) cited a number of conflicts in Paris, over Les Halles, La Défense, the wine entrepots at Paris–Bercy and the abattoirs at Vaugirard in the south of the city. Perhaps the largest and longest running controversy in the 1980s was that surrounding La Villette, a complex of disused abattoirs, warehouses and canals in the inner north east suburbs. Forty abattoirs, assorted skinning sheds, offices and markets on a site of 166 ha were already considered old fashioned by the 1950s, and some limited modernisation took place.

However, changes in wholesaling and retailing patterns, coupled with the trend towards slaughtering livestock near its place of origin signalled the closure of the main facilities in 1974. Subsequently, a number of redevelopments were proposed. From a land use point of view there were two difficulties to be overcome. One was the common planning issue of assembling a redevelopment site from a complex pattern of plot ownerships which covered both the public and private sectors, followed by work to clear and reclaim the land. The other issue was fought at a higher political and planning level and concerned the needs and sensitivity of the working class community which was already experiencing profound economic and social changes. This in turn was complicated by the intention of the socialist government which came to power in 1981 to stamp its mark upon the Parisian landscape.

Plans changed a number of times, and here there were close parallels with the London dockland development schemes, but a guiding principle throughout has been the need to provide a congested part of the city with more open space, better leisure facilites and a higher quality environment. On the abattoir site itself, a mixture of exhibition areas, a museum of science and industry, a circus, a technology area and two new residential districts totalling 22 ha have been developed. As with so many other redevelopment areas, including the numerous Garden Festival sites in British cities, it is not only the actual pattern of land use which has changed in detail, but also the density of development which has shifted to a lower level, reflecting modern planning ideas on the use of urban space. In the process, activities which were mainly concerned with production have given way to those concerned with consumption and this will have many implications for the areas involved.

CAUSES, EXPLANATIONS AND PROCESSES

It is clear from the foregoing that the land use patterns of western cities have undergone widespread changes in recent decades, both in terms of the continued outward expansion and the internal patterning and organisation of space. History will undoubtedly judge the past three decades to have been unusually important ones in the evolution of urban land use. To explain these changes fully would involve a comprehensive analysis of the myriad economic, social, political and individual behavioural changes which have restructured so many aspects of urban life. There is not space for that analysis here, but it is important to highlight some of the main processes which help to explain the evolving pattern.

Above all, it is important to stress the role of the processes which shape and change urban land use, for they provide some of the key elements in helping us to understand urban spatial structures. These spatial patterns represent some of the clearest manifestations of the broad economic and social structures within urban developed societies. Working in Toronto,

Bourne (1976) identified four main processes controlling urban land use change:

1 the extension of the urban edge, or suburbanisation,
2 the renewal of the central area,
3 the expansion of the infrastructure, especially transport and
4 the growth and decline of nucleations such as the removal of industrial areas from the inner city and the growth of institutional and recreational centres in the suburbs.

Although it is to some extent implicit in Bourne's findings, one might wish today to pay more attention to the broadly social and environmental issues which affect the quality of urban lifestyles.

One thing which does now seem fairly clear is that population growth per se is no longer a major reason for the expansion of the built up area of existing cities. The reason for this is quite simply that the majority of large cities in northern Europe and the urban heartland of north eastern USA have had populations which have been static or declining in recent years. Between 1971 and 1981, for example, population losses were recorded for nineteen out of the twenty largest urban areas in Great Britain (only Plymouth increased its population). A similar, although less marked, decline occurred in the USA. Out of 61 cities with more than 250,000 people within their municipal limits in 1980, 35 showed population losses over the previous decade. The population peak for many cities in the north east of the USA occurred in the 1950s and the aggregate population in cities over half a million in size was less in 1980 than it had been in 1960.

The major demographic impact upon land use in many cities has thus come not from the growth of population, but rather from its restructuring. It is mainly the social and age changes of the past generation which have contributed to the growing demand for new houses and other land using activities. The most powerful forces have been those affecting family and household size. In particular, the gradual break up of traditional patterns of family life through marital separation and divorce, the ageing of the population, a decrease in the number of children per family and a decrease in shared accommodation have all resulted in an increase in the number of separate households, and hence an increase in the number of dwellings needed. The population of Great Britain grew by just 0.57 per cent between 1971 and 1981, but the number of separate households grew by 7.4 per cent. This reflects a fall in average household size from 2.88 to 2.71 persons. For a city of 250,000 people this change alone would generate a need for 5,500 extra houses which would take 220 ha of land at typical suburban densities of 25 houses per ha. In the USA a similar fall in average family size occurred, from 3.33 in 1960 to 2.66 in 1987. In cases where the population has increased overall, urban areas have expanded by variable amounts, but the ratio has often exceeded unity. In Italy, for example, with particular problems of old and congested cities, every 1 per cent of population increase between 1961 and 1971 prompted an extension of the urban areas by 3.78 per cent – Table 4.14.

Coupled with demographic changes have come increased expectations over living standards which have resulted in greatly increased land needs. Much residential stock in European, and to a lesser extent in North American inner cities is old or obsolete. In the city of Stoke-on-Trent, for example, there remain approximately 24,000 houses dating from before 1919, despite the clearance of a similar number since 1945. Most of these consist of small terraced houses which lack both the full range of modern facilities and the room for expansion or improvement. Often it is the entire neighbourhood infrastructure, not simply the houses, which needs replacement. It has been calculated (Kivell 1975) that only 15 per cent of the 68,000 houses built in North Staffordshire between 1945 and 1972 were needed to cope with population growth, the far greater proportion making up the balance were required to accommodate the demographic changes and replacement needs noted above.

The planning styles, as well as the lifestyle requirements of the past few decades, above all the need to cater for growing motor car ownership, have resulted in most new housing development being undertaken at relatively low suburban densities. In Britain these have typically been 20–30 houses per ha compared with inner city densities from the beginning of the century which were commonly three or four times as high. Even the relatively high density inner city housing schemes undertaken by the public sector in the 1960s and 1970s were built at only half the density of the housing they replaced. Urban renewal in the cities of mainland Europe followed similar trends although there the emphasis upon flats has been greater and residential densities are generally higher than in Britain and North America. Finally, in the search for higher living standards, many urban land use needs have had to be satisfied well outside of city boundaries. This is especially true of such activities as recreation, airports, water supply, refuse disposal and mineral extraction.

In the employment sphere too there have been many changes with land use implications. Most notably, in the mature industrial cities of the western world, the rundown in manufacturing activity and the growth of service

Table 4.14 Expansion of urban areas for every 1% rise in population, 1961–71

Belgium	1.07	Luxembourg	2.14
Denmark	0.99	Netherlands	0.69
France	1.00	United Kingdom	2.09
Ireland	2.09	United States	0.77
Italy	3.78		

Source: Hauser 1982

sector jobs has already been alluded to. In the majority of cases this does not involve a simple switch of land from one sector to the other since the locational and status requirements of many service sector activities makes abandoned industrial land unattractive to them. Within the manufacturing sector the changes for Britain have been well documented (Fothergill *et al.* 1985; Spencer *et al.* 1986; Lever 1987). According to Fothergill *et al.* two trends have been important. First, there has been a decline in the number of workers per unit of floorspace, and, second, there has been a concentration of increases in the stock of manufacturing floorspace in small towns. Between 1967 and 1982 there was a marked thinning out of the density of industrial employment with the average number of workers per thousand square metres of floorspace falling from 36.0 to 21.4, with the falls being largest in London and the other main conurbations. Industry has changed from being labour intensive to being capital intensive, using more plant and land per unit of production in the process. Similar trends are responsible for the flattening of the employment gradient reported by Macdonald (1985) for Chicago. In 1956 the net employment density for manufacturing in Chicago declined by 14 per cent per mile from the CBD, but by 1970 the gradient had flattened to 11 per cent per mile.

In the city centres too, a number of processes have contributed to widespread land use changes. In northern Europe, including Britain, the spur for these changes was often the damage inflicted during the Second World War and the way in which it highlighted the need for extensive restructuring. Further boosts were given to central area redevelopment by the economic booms in the 1960s and 1980s. Essentially two things have happened. First, a number of traditional city centre activities have chosen, or been forced, to move out. Above all, residential land uses have been squeezed out by high land prices and lifestyle preferences, but industry, utilities and more recently some retailing and other commercial activities have also vacated the centre. Second, a number of planning strategies have been devised to make city centres more attractive and efficient for those activities which planners and the market deem to be prime central area users. These include the consolidation of office and specialised retail activities, and the improvement of transport and other infrastructure. What is perhaps remarkable about this, is that although locally there is considerable variation in detail, the broad processes may be observed across a wide range of urban situations from small, old established European market towns (Englestoft 1989), to large industrialised metropolises in Britain and America (McDonald 1985). Increased functional and land use specialisation in the core has been the result, but it has come about in two ways. Partly it is a product of the deliberate policy of segregating functions, as recommended by the Charter of Athens, but also improvements in the capacity, efficiency and accessibility of the city centre have induced rising land values which in turn have favoured high order office and retailing activities. In the American CBD the intensity of land use in manufacturing, transport, communications and utilities all declined between 1950 and 1970, but the commercial sector increased its intensity of usage (McDonald 1985). Bourne (1976) and Wilder (1985) both argued that distance from the CBD was important in determining the potential for land use change, but this argument becomes increasingly difficult to sustain as the structure of the city changes from monocentric to polycentric (Leven 1978; Muller 1981)

The cumulative effect of these processes, and the mechanism to which they can all be linked is that of decentralisation, and, to take it a stage further, that of counterurbanisation (Champion 1989). Central metropolitan areas have been declining in some ways for a generation and decentralisation during that period has fundamentally altered large urban structures. Office activities and specialised retailing have held on to their locations in the CBD longer than most activities and there are concerted attempts by many business communities to regenerate the core through new offices, convention centres, hotels, new retail forms and even some 'gentrified' housing. Today, however, especially in North American cities, even those activities considered quintessentially CBD uses are decentralising. Integrated complexes of offices, regional shopping malls and industrial parks emerge in the suburbs and are even reclustering in forms described as 'new downtowns' or 'suburban nucleations'. Wood (1988) gave the examples of Port America in Maryland, a planned CBD in a former tobacco field, and Crystal City, Arlington, Virginia, a new suburban complex of offices, shops, apartments and hotels, and argued that meanwhile the traditional inner city is being increasingly suburbanised. In Britain, tighter planning controls and higher land prices have prevented most such developments, but there are renewed signs that certain kinds of shopping are going out of town and offices are vacating London for business parks on the periphery or for provincial cities. In 1989, 7,000 office jobs were transferred out of London, including moves by the TSB to Birmingham, Pearl Assurance to Peterborough and sections of the Inland Revenue to Glasgow.

In mainland Europe the decline of manufacturing and warehousing in the inner city and the challenges to traditional retailing from out-of-town centres have all been similar, and there are many signs of decentralisation. However, there are differences. The European city is traditionally more compact, has better public transport and has retained a higher degree of middle class housing. The major land use components which characterise the European city are still recognisable, even though later phases of development have often obscured some of the earlier ones. Around remnant, congested medieval centres lie the grandiose planned developments of the eighteenth-century aristocracy, the industrial accretions of the nineteenth century and the suburbs, commercial centres, industrial estates and urban roads of the twentieth century.

As the present century winds to a close, there are signs that many commu-

nities are taking advantage of the lull in population growth to discuss what
kinds of cities and urban lifestyles they want and can afford. In particular,
there is some readiness to re-examine the rigid compartmentalisation or
zoning of land uses, with locationally separated areas of homogeneous
housing, industry, commerce or open space which has dominated planning
for forty years. Mixed land uses and functions, diverse styles, types and scales
of building, the renewal of old areas and the improvement of public transport
are all coming back on the planning agenda, with the encouragement of the
European Community's Green Paper on the Urban Environment. In particular, such ideas offer the possibilities of reducing unnecessary intraurban
movements and cutting down on traffic congestion. New concerns for the
quality of urban life, the protection of the environment and the needs of an
ageing, and perhaps more conservative, population are set to have significant
effects upon the pattern of urban land use during the next quarter of a
century. These themes will be returned to in the concluding chapter, but for
now it is sufficient to note that the cumulative effect of the changes is
producing substantially new settlement structures.

5

LAND OWNERSHIP

As with patterns of land use, the patterns of land ownership are important but poorly understood aspects of urban development. The principles and supposed effects of ownership are hotly debated from time to time but empirical evidence is thin. There is even justification for the claim that less is known about the pattern of land ownership in Britain today than at the time of the Domesday Survey nine centuries ago. Unlike some of its continental neighbours, Britain does not have a comprehensive cadastral survey and register.

Ownership, where land is concerned, is a far from simple concept. With long and varied histories, most European countries have evolved complex patterns of land holdings and tenurial rights. For Britain, Denman (1978: 101) suggested that 'it would be exceedingly difficult to identify and classify all tenurial systems in existence'. What is clear is that land tenure involves a complicated collection of rights to own, occupy, use or improve space and to lease, sell or pass it on to one's heirs. It consists in part of physical attributes such as size, topography, location and accessibility, and, for the other part, a set of institutional and legal rights and obligations. These latter are essentially social constructs which vary from country to country, and from time to time. According to Ratcliffe (1976: 21) 'systems of land tenure embody those legal contractural or customary arrangements, whereby individuals or organisations gain access to social or economic opportunities through land ... land without the dimension of tenure is a meaningless concept'.

The principal forms of land holding for Britain are freehold, with either a private individual, or a corporate or state body possessing outright ownership, and leasehold where a tenant leases the land from its outright owner. Conditions governing leaseholds of land, and the buildings which may stand upon it, are so numerous as to be almost infinite. Leases may themselves be sold, or sub-let and the length may vary from a few months to 999 years. In the latter case, the leaseholder may have almost as much effective security of tenure as a freeholder, but will be constrained by regular rent reviews and other conditions. In Britain, relatively long leases are common; for commercial property 25 years is the norm (with periodic rent reviews), whereas in

93

mainland Europe and the USA, 9 or 10 year leases are typical. During the process of urban development an individual plot of land may pass through a complicated sequence of ownerships and leases. Depending upon the speed of development, the location of the land relative to the urban area and the speculative behaviour of the participants, this state of flux may last from a few months to a couple of decades.

THE IMPORTANCE OF LAND OWNERSHIP

The notion of land ownership is not merely an issue of arcane legal debate, it has implications of great importance for urban development which can be summarised as follows.

1) The size and configuration of land holdings profoundly affects urban morphology. The layout and scale of urban development owes much to the nature of original land ownership boundaries (Conzen 1960; Ward 1962; Dyos 1968; Mortimore 1969), and the reconstruction of extensively damaged or blighted areas is often constrained by the original pattern of plot ownership. Many of the open spaces in Europe's older cities exist today because of the conservationist nature of their orginal owners, or because of the protection afforded by common ownership. At the very least, the pattern of original ownership can still be traced in many cities from the evidence of street names.

2) The timing of land sales affects the nature of urban development. In particular, this may reflect the contemporary technology and economic driving forces together with architectural and planning styles. Railway era housing, for example, differs from that of the motor car age, and the industrial or resort town of the nineteenth century is very different from its twentieth century counterpart.

3) Land ownership confers power, indeed until the mid eighteenth century in England it was the very cornerstone of power and the big land owners were the economic and political leaders of society. Today a few of the traditional land owning families still feature amongst the wealthy in most West European countries, but their power is far less than previously. There is however a new generation of hugely wealthy financial corporations, frequently international in scale, with extensive land owning interests. At the other end of the spectrum, millions of individuals, representing the majority of households in most western nations, have gained considerable financial power and flexibility from owning their own houses and the small plots of land which they stand upon.

4) Landowners may exert considerable influence over urban planning policies, especially if they act in concert (Cox 1984). This comes about through their decisions on whether, or when, to sell land and participate in different kinds of development. Adams *et al.* (1988) identified an important distinction between active and passive land ownership, and the way in which

94

this affected development. In addition, land owners have influence over the preparation and execution of land use plans. In local plan areas in Cambridge, Greenwich and Surrey Heath, Adams and May (1990) found that one-quarter of all representations by land owners resulted in a subsequent change in the plan.

Britain's planning system, like many others, is largely negative, i.e., it can prevent or modify proposed developments, but it cannot force private sector land owners and developers to undertake a particular scheme. In market economies there is competition over access to land and property, rent levels and conditions of occupancy between land owners, production interests and consumption interests. According to some commentators (Montgomery 1987), it is this process, not planning intervention, which is mainly responsible for allocating land to different users.

5) Land ownership is an integral part of both national and local economies and it can be seen as a part of the relationship between the production sector and the consumption sector. The former sector views land as a commodity and comprises developers, together with farmers (in one capacity) and speculative owners whose main interest is to maximise the exchange value of the land. The latter sector consists of those who occupy land for a specific purpose, e.g. industrialists, retail and office companies, home owners and farmers (as agriculturalists), whose main interest is to maximise the use value of the land. In addition to being a tradeable commodity, land can be used as a reliable asset against which to raise loans, for example, for business expansion. Finally, the role of land in national economies can be seen in terms of employment. The land and property industry *per se* is highly capital intensive (see Table 5.2), but substantial numbers of jobs exist in related activities, for example, in planning, surveying, valuation, estate agency and the building industry. The building industry in particular is one of the most significant indicators of change in the state of the economy.

6) Finally, a consideration of land ownership is important for what it reveals about the nature of society, given that ownership is a social construct. Across the spectrum from free market economies such as those of the USA and Japan, through the mixed economies of much of Western Europe to the centrally planned regimes of the Eastern Bloc, it is the ownership and trading of land which is a key characteristic of the differing societies. In the rapidly changing economies of Eastern Europe today, debates about the ownership of land and property are playing an important role. In many countries, notably the USA, Japan and Britain, the ownership of land is deeply etched both in the national culture and in the individual psyche, and in Australia, Slaughter (1973) suggested that 'love of ownership is inbred'.

WHO OWNS LAND?

As already suggested, our knowledge of urban land ownership is woefully limited. In Britain the problem is especially acute. Broad patterns are known, for example, the decline of the traditional land owning estates, the shift towards home ownership, the existence of significant public sector land holdings and the rise of major property companies and property divisions within insurance companies and pension funds, but there has been a marked lack of empirical study of land ownership. Only in a few specialised areas of concern, such as derelict or vacant land on the public sector land registers are systemmatic figures available.

This lack of information on land ownership has proved to be a fundamental problem affecting many aspects of land policy, a point that has been identified by such diverse writers as Denman (1974: 46), Massey and Catalano (1978: 4), Flatt (1982: 329), Norton-Taylor (1982) and Goodchild and Munton (1985). Even where information has been collected it is rarely comprehensive, wholly reliable or freely available. Shortcomings are identified for example by Harrison, Tranter and Gibbs (1977: 14):

> All the studies yet made of land ownership have been restricted almost entirely to the establishment of elementary facts. Their basic statistical coverage has varied widely, both in terms of the samples employed and the categorisation of ownership adopted which has nowhere begun to match the complexity occurring in practice. Consequently almost nothing can be concluded on which normative and policy making decisions can be based.

This observation is based upon a study of agricultral land and even these 'elementary facts' rarely exist in urban areas. In Britain there are many professions with a detailed but localised and incomplete knowledge of land transactions, but 'such material is rarely published in any systematic form, as opposed to anecdotal form' (Barrett and Healey 1985: 12).

Two reasons go a long way towards explaining the shortage of data. First, there exists in England a tradition of confidentiality over land ownership; transactions tend to be 'exclusive and confidential' (Edwards and Lovatt 1980: 3). The detailed cadastral surveys available in other European countries have no counterpart in Britain, and the limited and incomplete nature of the records of HM Land Registry, which was not open to the public until 1990, have hampered research.

Second, although local authorities, and other public bodies, collect a great deal of information on the use, development and ownership of land (mainly their own), most of this is undertaken in a very fragmented and ad hoc manner. Most property information systems to date have been unique, and were designed to undertake specific tasks. The, perhaps surprising, consequence of this is that the majority of local authorities are unable, quickly and accurately, to identify their own land holding positions, and are completely unequipped to identify the ownership of privately held land in their area. It is to be hoped that, for the public sector at least, the new generation of computer based property information systems will lead to the facility for straightforward measurement and simple comparisons.

Similar problems, although often in a less acute form, affect the picture of urban land ownership in most countries. Given the undoubted importance of land ownership for the patterning of metropolitan activities, these short-comings are regrettable. Perhaps, in view of this, it is not so surprising that supply side considerations and land ownership patterns have been so neglected in the neoclassical economic models of the city.

For a better understanding of the overall pattern and significance of urban land ownership, it will be useful to consider the private and public sectors separately, and then refer briefly to the trend towards partnerships between them.

THE PRIVATE SECTOR

In all of the developed western nations there exists a strong legal and social right for individuals, companies and other private sector bodies to own land. These rights are jealously guarded, but they are rarely absolute, being constrained by a variety of state legislation. In general, private property rights may be limited by:

1   The exclusion of certain social groups from ownership.
2   Restrictions on the use and development of land according to planning or zoning laws.
3   Taxation of land itself, its beneficial use or betterment.
4   Expropriation of land by the state.

Even the most ardent supporters of free market principles and private owner-ship usually accept that in densely settled urban areas some measure of public intervention is necessary in order to avoid land use conflicts, to protect the environment and to provide basic infrastructure. Beyond that, however, the nature and extent of intervention is disputed.

Exactly how much land is in private hands, and how much wealth it represents is almost impossible to calculate. Fleming and Little (1975) esti-mated that, in Britain, land per se was a very small part (4 per cent) of personal wealth, but their figure excluded housing. Certainly there is some evidence to suggest that its ownership is very concentrated. Norton-Taylor (1982), for example, estimated that 84 per cent of Britain's land was owned by 9 per cent of the population. Once again, these figures are questionable. They are based upon area, not value and they do not distinguish between urban and other land. Like most such estimates the figures quoted take no account of the diffuse and collective pattern of private ownership represented by individual interests in pension funds, insurance firms and other publicly quoted companies.

The structure of private ownership is complicated because there are so many different forms it may take and because, in the process of urban development, land may pass through many hands. For this latter reason, and because development is rarely carried out by the original owner, Goodchild and Munton (1985) use the term 'pre-development land owner', but this is too restrictive for present purposes. Cox (1984) identified the following groups in the development process, all of whom may at some stage actually own land: 1) land owners, including the crown, financial institutions, farmers, public and private companies; 2) the rental sector, property companies and developers; 3) the financial sector, banks and insurance companies; 4) the commercial sector, shops and offices; 5) the productive sector, farmers, manufacturers and others producing commodities for sale. It is obvious that there are some overlaps, for example, farmers may belong to both the productive and the land owning sectors, and composite holding companies may have activities embracing more than one of these headings.

A more direct typology of land owners is the threefold grouping into former landed property, industrial ownership and financial ownership as

Table 5.1 Major urban land ownership groups: some basic characteristics

Land owner	Predominant types of land and land use	Main activity
Former landed property		
Church	Residential low income	Rent
	Residential high income	Develop
	Offices	Develop
Landed aristocracy	Residential low income	Rent
	Residential high income	Rent/develop
	Offices/shops	Develop
Crown Estate	Residential low income	Rent
	Residential high income	Rent/develop
	Offices	Develop
Industrial land ownership		
Manufacturing industry	Industrial	Owner occupation
(construction companies)	Office/shop	Develop
	Residential	Sale
Financial land ownership		
Financial institutions	Offices	Develop/invest
	Industrial	Develop
Property companies	Residential	Rent
	Offices/shops	Develop/invest
	Industrial	Develop

Source: adapted from Massey and Catalano 1978

proposed by Massey and Catalano (1978), Table 5.1. This will form the basis of discussion for the rest of this section; however, for the sake of completeness, a fourth category will be added, that of home owners. Although this group is not involved in land as a business, it is numerically and spatially important in most western cities and it does play a major role in urban land markets.

Former landed property

This is a group which, in the developed world, is largely confined to Europe, although British landed interests were also active in North America and Australia in the nineteenth century. The estates of the landed gentry, aristocracy, church and crown all predate industrial capitalism and were often assembled through judicious marriages or military and commercial ventures.

The majority of their land was, and still is, agricultural, but some members of this group, however, did build up valuable urban estates. The widespread urban growth of the eighteenth and nineteenth centuries rocketed a few of them into the realms of the super-rich (Sutherland 1988), but they were not simply passive beneficiaries of this process. Indeed a number of them were very active and astute developers. In London it was the Grosvenor (Dukes of Westminster), Howard de Walden (Dukes of Portland), Portman and Cadogan families who were most powerful, and their estates remain largely intact today as a recognisable imprint on the London landscape. Only the Bedford estate has all but vanished. Other towns also possess large, old established family estates. Among them are Sheffield (Duke of Norfolk), Eastbourne (Duke of Devonshire), Chester and Liverpool (Duke of Westminster), Norwich (Coleman family), and in Lanarkshire the Duke of Hamilton has large urban holdings. In mainland Europe too a number of private owners have held on to large estates, for example, the Thurn und Taxis family in Germany. In Britain, apart from individual families it is the church commissioners and the crown which remain the largest representatives of this group, although some of the older universities are also major urban land owners. Most of the members of this group were large housing landlords in the nineteenth century city, but it is commercial land which has proved the more attractive holding in the twentieth century.

Cox (1984) argued that landed interests have been involved in a successful social, economic and political holding operation since the nineteenth century. They still have considerable political influence and can use an economic veto, or delay, on development by refusing to sell land; they can also affect planning legislation through support for green belts and other urban containment measures. He suggested that the land owning classes, together with the crown, churches and universities still own about 40 per cent of the land in Britain, but again this is by area, and takes no account of high land values in urban districts where they are far less well represented. Although

99

there are increasing pressures upon all land owners to manage their holdings with economic efficiency, it is clear that many of them take account of tradition, sentiment and social obligations. Sutherland (1988) concludes that the record of the traditional large urban land owners has been a good one.

Industrial land owners

Under this heading, land ownership may be seen as as a necessity for production. It covers farmers, manufacturers, and a variety of commercial interests. Land is held primarily in pursuance of these activities, but it also enters the balance sheet as a cost or an asset, and it may be bought, sold, leased or mortgaged to facilitate business expansion or to raise capital. In this context, land ownership, values and the behaviour of owners may be important in determining the use of land and hence urban patterns. Old established uses, such as manufacturing may be squeezed out by a more profitable use (such as offices), and a barrier may be imposed to the entry of uses of low monetary value such as community services and low rental housing (Montgomery 1987).

This group is generally opposed to interventionist land policies, but it is divided in its interests, for example, what suits the manufacturer in terms of land policy and taxation, may not suit the farmer. Industrialists have been closely involved in the recent restructuring of urban economies and the run down of the inner city, and their land needs, both in terms of their locational requirements and their ability to outbid other activities have changed profoundly. They, and others in the productive sector have important influence on policy, and indeed powers of veto, through their willingness or unwillingness to invest. For this reason, the past decade has seen the planning and land policies in a number of countries selectively softened in order to encourage urban development or redevelopment.

Building and construction companies form the final element of this group, but they do not sit comfortably within it, and are themselves a diverse collection of interests. Building is a long term business in which companies need the security of a reliable supply of land at predictable prices. For this reason, many firms hold land banks of sites acquired perhaps some years in advance of their needs. At times they may sell surplus land, thus involving themselves in the financial category (below). Such speculative behaviour has been criticised from time to time, particularly with respect to the price and availability of land for house building. Claims that unremitting rises occur in the price of land, and that completely unearned and risk free gains are to be made in this way have, however, been disputed. Studies in Britain and Germany (Hallett 1977) have shown that the price of building land can go down in real terms as well as up. It fell sharply in the mid 1970s, and at the time of writing a number of major building companies, including Costain and AMEC have written down their asset value because of the falling price of land.

The building and construction industry is normally involved in piecemeal urban development, but recently in Britain there have been proposals for private sector new towns built entirely from scratch (Lock 1989). These plans are the product of large companies working together to provide new settlements of around 10–15,000 people, in congested areas such as the South East of England. A number of test cases have been rejected on planning grounds (e.g. Tillingham Hall in Essex and Foxley Wood in Hampshire), but one, at Brenthall Park, was approved in 1987.

The response of the productive sector to government intervention in land is mixed, but generally negative. Some elements, notably the farming and building industries maintain strong political pressure groups to promote their causes. The large general construction companies may be more likely than small speculative house builders, to accept limited intervention in the form of betterment levies and land use controls, but they remain united in their opposition to widespread public land ownership.

Financial land ownership

The financial sector is the most concentrated in urban areas and it is, in many ways, the most controversial. It consists principally of banks, finance houses, insurance companies, property development companies and pension funds. This category deals in land almost wholly as a commodity which can generate capital gains, rental income and investment security. The political power of the sector is limited by its fragmentation, its secretive nature and internal competition, but it has two powerful forms of economic influence:

1   by controlling vast amounts of investment capital it may influence financial markets; in particular, by limiting the finance available to the productive sector or by moving investment from one region, or country, to another; and

2   any state policy which undermines confidence in the property sector may upset financial markets and perhaps precipitate an investment famine; governments thus tend to treat it with respect.

There are many ways in which the financial sector holds land, but the main activities comprise direct freehold purchase, leasehold purchase, sale and lease-back arrangements, joint companies with developers, investment in development companies and mortgages secured against property. It is also important to recognise that there is not one single property market, but several. Thus office, industrial, retail and housing land often operate differently and experience different degrees and timings of slump and boom conditions.

It was after the Second World War that large property and development companies began to make a major impression upon the urban face of Europe and North America. In Britain, rent controls and other restrictions, together with large scale public sector activity made residential property relatively unattractive as an investment. At the same time there was a revolution in the scale and organisation of both retailing and office activities, so property companies concentrated upon these sectors. By the mid 1950s the developments pioneered in North America were spreading throughout Europe. The central areas of London, and most provincial cities, were profoundly remodelled between 1955 and 1975, although the uniformity of building materials and styles, and the promotion of corporate images often resulted in a monotonous sameness. Few companies had the financial and technical resources to undertake such large scale developments so the process became concentrated in the hands of a small number of developers.

Most of the development was financed by the growing funds of various institutions, notably clearing banks, merchant banks, insurance companies and pension funds which provide longer term investment finance. The latter typically keep between 10 and 20 per cent of their assets in land and property, with a distinct preference for blue chip urban property. The BP pension fund, for example, had a property portfolio valued at £635 million in 1987 (Sutherland 1988), including much of London's Berkeley Square. Pension funds and insurance companies are far and away the largest investors in Britain's urban land and property but during the mid 1980s they began to decrease their property investments, Table 5.2, in favour of the equity market, and to some extent the banks, especially overseas banks, moved into the market.

Some of these property companies are very large, with financial values that exceed those of the biggest industrial companies in many western nations. Table 5.3 shows a sample of the largest British based companies, but it should be viewed with caution because the 1980s represented one of rapid change in this sector, because the figures for any single year may be distorted by a single large project, and because by 1992 the property sector was severely depressed. In May 1992, Mountleigh collapsed with debts of £500 million.

Table 5.2 Net property investment by selected institutions

	1982	1983	1984	1985	1986	1987	1988	1989[a]
				(£ millions)				
Insurance companies	1,059	845	744	815	845	842	1,411	757
Pension funds	983	680	906	509	379	141	238	22
Property unit trusts	57	−9	47	−5	−101	−515	93	29
Others	152	114	132	183	174	−69	258	–
Total	2,251	1,630	1,829	1,502	1,297	399	2,000	808

Note: [a]First quarter only.
Source: Cadman, D. and Austin-Crowe, L. (1990) *Property Development*, third edition (ed.) R. Topping and M. Avis, London, E. and F.N. Spon. Their source: Department of Trade and Industry, *Money into Property*, Debenham, Tewson and Chinnocks

Table 5.3 Selected large property companies in Britain[a]

Company	Sales (£000)	Profit before tax (£000)	Employees
Mountleigh	529,000	70.7	112
MEPC	217,000	104.8	886
British Land	150,800	56.4	479
Rosehaugh	78,700	30.3	215
Speyhawk	77,800	17.3	497
Capital & Counties	73,100	43.5	583

Note: [a]Figures relate to financial years ending in 1988.
Source: Property Development, Keynote Publications, Hampton, Middx

It has been argued that there are certain adverse effects in this concentration of ownership and development (Ambrose and Colenutt 1975). In particular, the effects on local employment, living and housing costs, the over-rapid transformation of social and employment structures, the creation of wealth inequality and a poor balance of investment which disadvantages housing and manufacturing are cited. These criticisms have some validity, but they can also be seen to be based upon an underestimation of the post-industrial transformation of advanced urban economies. It is also true that most of the assets of the insurance companies and pension funds which own these large developments, are held in trust for the millions of individual policy holders and beneficiaries.

The pattern of ownership is thus less concentrated than it may appear. Even so, a number of problems arose towards the end of the 1980s as the boom pushed up land prices and distorted the market with an oversupply of retail and office space and a shortage of land for industrial and warehousing activities (*Financial Times*, 15/6/90). In 1988 there were between 5 and 6 million square metres of additional office floorspace in progress in London, but by 1990 the office activity boom was beginning to slacken and developers were finding it difficult to let all of the space. In the longer term, the success of large scale office developments in London depend upon that city's ability to consolidate its position as the financial capital of Europe.

Despite the apparent security of land and property as an investment, the 1980s demonstrated what a volatile sector it can be. As the economy pulled out of the recession of the late 1970s and early 1980s, and as planning regulations were relaxed to stimulate development, virtually all property and land companies experienced rapid expansion. In addition to office growth which was stimulated by the deregulation of the City and the spread of white collar occupations generally, there was extensive growth of shopping centres, home ownership and hotels, prompted by a consumer spending boom. A number of public bodies took advantage of the boom by selling off surplus land; British Rail, for example, made £319 million in 1989–90 from land sales and leases. By 1990, however, high interest rates and flagging business confidence forced a slowdown and a number of smaller property companies found that their rental income no longer covered even their interest payments. Many development schemes were quietly abandoned, postponed or cut back.

During the 1980s, development activity in London was concentrated into relatively few hands. In particular three developers, and their companies, have dominated the largest schemes, including those at Broadgate (Liverpool Street), King's Cross and Canary Wharf in the Docklands. They are Godfrey Bradman, Stuart Lipton and Paul Reichman, with their companies Rosehaugh, Stanhope, and Olympia and York. Between them they are having a greater impact upon London's built environment than anybody since John Nash (Knobel 1988). Olympia and York, based in Canada, has major holdings in Toronto, Boston, San Francisco and the World Financial Center in Manhattan, although even companies of this size are not immune to the effects of a slowdown in development activity such as that of the early 1990s. In May 1991, the combined effects of a business recession, falling rents and an oversupply of office space in London, resulted in the company postponing further development at the partially completed Canary Wharf complex, Europe's largest office development. In May 1992 Olympia and York filed for relief from creditors under both Canadian and US bankruptcy laws.

The development industry has been becoming increasingly international for nearly three decades. In the case of Britain a slowdown in rental return and a tightening of competition in London in the mid 1960s caused domestic development companies to look first to provincial cities and then overseas, notably to Europe, North America and Australia. Elsewhere, the accumulation of wealth in the Middle East and a recent relaxation of financial regulations in such countries as Japan and Sweden, have encouraged these countries to become major players in a worldwide property and development industry.

In the late 1980s, the London property market saw heavy investment from overseas. Between 1985 and 1989, foreign investment in central London increased tenfold (*Financial Times*, 31/8/90). The Japanese were particularly active and Britain became Japan's main target within Europe, attracting investment of $4,631 million in 1989, compared with $1,577 million in France, $182 million in Spain and $66 million in Germany (*Estates Gazette*, 9022, 1990). In 1989, Japan was responsible for 43 per cent of all overseas investment in London, but Scandinavian and European Community interests were also strongly represented. This Japanese interest is a reflection of limited opportunities, high competition and the introduction of capital gains taxes in their domestic market. The Tokyo stock market is underpinned by exceptionally high land values which saw particularly rapid increases in the 1980s – Figure 5.1. Many analysts feel that land and property prices there are artificially high and there are now moves to reform land laws and taxes.

In the USA and Canada too, the Japanese have been particularly active,

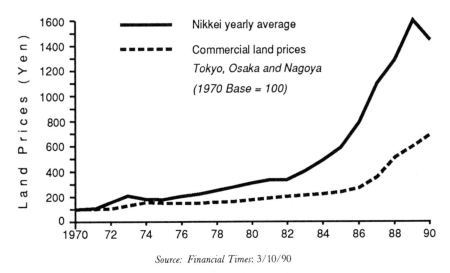

Source: Financial Times: 3/10/90

Figure 5.1 Commercial land prices: Japan, 1970–90

driven by the growth of their economy, the strength of the yen and a high national savings rate. Japan is now the largest foreign landlord in the USA (Wharf 1988). By 1987, Japanese firms owned approximately 9 million square feet of office, commercial and hotel floorspace in New York alone, Table 5.4, even the Rockefeller Center is now owned by the Mitsubishi Corporation. Almost one-third of downtown Los Angeles is owned by the Japanese and they are increasingly buying houses on the West Coast. To date this investment has not been unwelcome, and nearly every US state maintains a liaison office in Tokyo.

It is Australian cities, however, which most clearly demonstrate the internationalisation of land and property investment. Since the mid 1960s, British investment has been significant, but the 1980s saw foreign involvement raised to a massive new level, especially by Japanese and other South East Asian investors. Urban land in Australia is largely in private hands, but, as elsewhere, the distribution is uneven, and a small number of insurance and finance companies play a dominant role – Table 5.5. Important questions have been raised about the domination of Australian cities by finance capital, especially in terms of links between organisations and the representation of property interests on city councils (Kilmartin and Thorns 1978). By 1973, 14.2 per cent of Sydney's CBD buildings were foreign owned (Adrian and Simpson 1986) and in 1976 a Foreign Investment Review Board was established to control overseas investment. After 1980, a new wave of investment flowed into Australian cities from Japan, Hong Kong and Singapore. This took two forms: relatively small investments in residential property which do not show up in official statistics, and large investments in commercial real

Table 5.4 Japanese real estate holdings in New York region, 1987

Investor	Sq ft	Floors (N)	Ass'd value ($millions)	Location
Office/commercial				
Mitsui Fudosan	2,200,000	54	610	Midtown[a]
Kumagai Gumi	1,500,000	60	384	Lower Man.[b]
Kato Kagaku	600,000	44	301	Midtown
Hiro Real Estate	540,000	38	250	Midtown
Sumitomo Corp.	493,860	42	80	Lower Man.
Shuwa Corp.	480,000	40	175	Lower Man.
Kowa Realty	420,000	35	60	Midtown
Hiro Enterprises	343,160	23	29	Lower Man.
Sumitomo Life In.	270,000	36	20	Midtown
Hirokohi	267,000	27	105	Midtown
Mitsui Fudosan	225,000	28	18	Lower Man.
K. Hattori Seiko	200,000	11	7	Mt Olive, NJ
Nippon Life	164,000	22	8	Midtown
Kenwood Electr.	120,000	9	8	Mt Olive, NJ
Dai Ichi Real Est.	94,000	9	94	Lower Man.
Pacific Real Est.	70,000	12	19	Midtown
Hotels				
Nikko/Japan Air.	724,000	38	175	Midtown
Caesar Park Int'l	400,000	18	29	Midtown
Tsuguto Kitano	96,000	12	21	Midtown

Source: Wharf 1988
Notes: Data obtained from field surveys and real-estate literature.
[a]Midtown is area on Manhattan between Canal St and 59th St (Central Park South).
[b]Lower Man. is area on Manhattan south of Canal St.

estate (Thrift 1986). South East Asian investors see Australia as a secure, long term prospect and are concentrating on CBD developments and multi-purpose office, tourism and residential developments such as the Chia project at South Yarra.

It is the state capitals, especially Sydney, which have been the focus of interest, and this infusion of foreign capital has been largely responsible for the rise of Sydney as the major financial centre, at the expense of Melbourne (Daly 1984). Perhaps the ultimate in overseas investment is the recent proposal by the Japanese Ministry of International Trade and Industry, to build a new city somewhere in Australia (Self 1988). This multi-functional polis with resort, convention, research and residential functions would be a private sector investment, but it clearly raises problems of an anti-Japanese backlash, and the prospect of Australian technology being overwhelmed.

Table 5.5 Land ownership in Auckland and Melbourne central business streets[a]

Land owner	Auckland (% total area)	Melbourne (% total area)
Public bodies	18.6	26.1
Financial institutions	27.1	28.9
Companies	24.1	8.7
Property development companies	4.5	18.8
Religious groups	15.4	8.7
Private trusts	3.8	0.8
Private individuals	1.7	3.8
Trade unions	0.9	–
Others	3.9	4.3

Source: Kilmartin and Thorns 1978

Note: [a]The central business streets used for the analysis refer, in the case of Auckland, to Queen Street from Karangahape Road down to Quay Street; and for Melbourne, the block bounded by Collins and Bourke Streets, including Little Collins Street. Source: *Melbourne Cityscope* and Valuation Department, Auckland City Council.

Home owners

The fourth major category of private sector land owners consists of owner-occupiers of residential housing. Although neglected by Massey and Catalano, it is important because it is a large and growing category which represents a major element of the land in most cities, and because it qualifies the apparently concentrated pattern of land ownership represented by the three previous groups.

Home ownership has grown rapidly during the past decade. The developed nations can be grouped into two categories: those where home ownership ranges between half and three-quarters of all households, including Ireland, Spain, the USA, Finland, Canada, Belgium, Norway, Italy and the UK; and those where it is between a quarter and a half, including Austria, West Germany, The Netherlands, Sweden and Switzerland. In this second group, in particular, leasehold arrangements mean that the households do not necessarily own the land upon which their houses stand. In the UK approximately two-thirds of households own their dwellings. In 1988 the amount outstanding in the form of mortgages and other housing loans totalled £224,236 million. Whichever way this is viewed, it constitutes a very substantial amount of property. Although building society loans made up two-thirds of the value of the typical house, the freehold was possessed by the owner occupier, so ownership is firmly in the hands of private individuals. Calculating exactly how much land is involved in this pattern of home ownership is not easy, but some crude estimates can be made. The number of owner occupied houses in different cities can be ascertained from the 1981 census. It is known that approximately one-third of Britain's housing stock is

107

pre-1919. Assuming a net residential density of 100 units per hectare for this, and 30 per hectare for the newer stock, estimate of the area covered by owner occupied housing can be made. For example, Plymouth had 87,100 owner occupied houses in 1981; using the assumptions above, these are calculated to cover a net area of 1223 ha, or 15.4 per cent of the city's total land. Similar calculations for other British cities suggests figures between 10 per cent for older, high density urban areas with relatively low levels of owner occupation, to rather more than 25 per cent for wealthier, residential towns.

Monopoly ownership

Having outlined the four major groups of land owners within the private sector, one further point needs to be addressed. This concerns the economic power of land owners and the extent to which they may use monopoly ownership to withold land from sale, hence driving up land prices and generally controlling the market. This is an argument which has frequently been used to 'explain' high house prices and to justify taking land into public ownership.

In fact, it is not easy for land owners to force prices upwards in this way. Certainly, they can wait for favourable prices before they dispose of their land, but there is little evidence to show that they can control the market by acting in concert. Supply constraints may be locally significant and the total amount of land is fixed; but the supply is not, in fact, fixed around a given urban area. As the town expands, new areas provide competition both for old areas and for each other. There is some localised evidence from Britain in the 1980s that builders were forced to make inflated bids for land which was in short supply, but in such cases it was normally the restrictive effects of planning, not monopoly behaviour by land owners, which was the root cause (White 1986; Evans 1988).

Land owners are not the tightly knit collective that they might appear to be. Each owner has his/her own strategy and aims, and times of rising prices are accompanied by intense competition. In an in-depth treatment of the subject, Goodchild and Munton (1985), concluded that, in the long term, land owners can theoretically affect the price by witholding their land from sale, but that this remains to be proved in practice. Detailed studies in Toronto, during a time of rapidly rising prices, similarly concluded that there was no evidence of land owners possessing monopoly power in this way (Markusen and Scheffman 1977). The concentration of ownership in the Toronto region was far too low to imply market power. Instead, the explanation for rising prices was firmly rooted in a combination of a sudden shift of financial assets from other, poorly performing, investments, strong population growth, red tape in the planning process and a lack of infrastructure. Taking a rather broader perspective in the UK, Massey and Catalano (1978: 186) concluded that 'there is no single group, based on land ownership by capital, which can be said to be a distinct and coherent fraction'. The main groups differ in their relationship to land, its place in the structure of accumulation and in terms of ideological and political bases. There is also little real coherence either within or between the groups.

THE PUBLIC SECTOR

Public land ownership has traditionally been justified for reasons of 'the common good' or 'the public interest'. These ideas found ready acceptance in many quarters in Britain during the immediate postwar period when comprehensive town planning was introduced, but they have always had their critics. Since the late 1960s, it has become increasingly clear that the concept of a single 'public good' is severely handicapped by the multiplicity of interest groups which exist in modern society (Meyerson and Banfield 1964; Simmie 1978).

More specifically, a large number of individual advantages have been claimed for the taking of land into public ownership, especially during urban development. Effectively, these can be condensed into three main arguments: planning efficiency, fiscal and social equity and the provision of services.

The planning efficiency argument, which has been discussed by Hall (1976) and Roberts (1977), inter alia, suggests that where governments or local authorities own the land required for development they can promote efficient and desirable land use patterns and channel growth in a rational and well co-ordinated manner. Cost arguments also come into it, and it has been claimed that public sector interest rates, carrying costs and servicing charges are lower. Kehoe et al. (1976) added to this the assumption that public ownership will eliminate delays in the land use regulatory system, and Shoup (1983) pointed to the way in which advance public purchase of land for development can ensure the preservation of the best sites for public facilities as well as a favourable purchase price. The final piece in this jigsaw is the suggestion that because a municipality has both a comprehensive overview of its own needs, and ultimate planning control over its own development, it will possess better information than the market about its long term land requirements. It will thus be able to internalise some of the conflicts (Montgomery 1987).

The argument that public land ownership can be used to achieve financial and social equity can be put forward at a number of different levels. At the broadest level it can be advanced as a part of the process of wealth redistribution. Rather more specifically, it has often been suggested that taking land into public ownership is a means of reducing both the inequity between land owners who do, or do not, receive development permission, and for ensuring that the community gains the overall financial benefit. This latter view is based upon the assumption that it is society which creates enhanced land values, and that the economic gain should therefore rest with the whole community and not, fortuitously, with the owner alone. This, of course, is a development of the classic view of J.S. Mill that the ordinary progress of a society which increases in wealth is at all times to augment the income of landlords: they grow richer, as it were in their sleep, without working, risking or economising. This view is not without its critics: Denman (1978: 91) for example, argued that land is little different from any other commodity and that economic progress and social conditions come about not by some impersonal progress of society, but by the positive, active decisions made by farmers, shopkeepers and householders. Either way, public ownership is not strictly necessary in order to recover enhanced land values for the community: this can be achieved through betterment levies or taxes.

The third argument in favour of taking land into public ownership is that it is necessary in order to allow public bodies, especially local authorities, to perform their primary tasks of providing houses, schools, hospitals, roads and other community services. The process of municipal land acquisition culminated in Britain in the period 1959–75, when local authorities made extensive use of housing and planning legislation. In explaining the pattern of publicly held land in most western cities it is these functions and powers which are of greatest importance. Until recently this was a largely uncontroversial sphere of public land acquisition, but with a variety of moves towards the privatisation of services there may be a diminishing role for public bodies.

Ranged against the arguments for public land ownership, are a number of counterclaims, and these, generally promoted by adherents of the free market, were in the ascendent during the 1980s (Lloyd 1989). Once again, for convenience, they can be condensed into three issues, viz. bureaucratic inefficiency, private rights and land values.

The bureaucratic inefficiency argument recognises that, even in the absence of a market, decisions about land use and development have to be made, but it questions the ability of government or local authority bureaucracies to produce clear decision making or satisfactory results. There is no evidence that the power play between bureaucratic segments, works particularly well (Bryant 1976), and Clawson (1971) felt that a public monopoly would be under a strong temptation to fall into unprogressive, insensitive and inefficient ways. Those British cities where a high level of public land ownership and a municipal monopoly over development were the norm for a generation do little to dispel these suggestions. Even in mixed economies where public and private ownership exist side by side, excessive public intervention is often blamed for distorting the market and causing delays in development and higher land prices (Markusen and Scheffman 1977; Nowlan 1977; Lloyd 1989). At the same time, it must be recognised that private land owners produce their own inefficiencies by sometimes sitting on land and not allowing others to develop it. A related claim against public ownership is that it is largely unresponsive to the knowledge and discipline of market supply and demand. Hamilton and Baxter (1977) argued that there is no evidence that public authorities could provide land better or cheaper than the private sector, and that they also possess the serious possibility of a major collective error.

Private property and the workings of a free enterprise society are thought by some to be threatened by public land ownership. A complete public monopoly over the ownership of land and the granting of planning permission, together with public sector use of the land in question is seen as a dangerous combination of power.

The third argument against public sector land holding concerns its supposed effect upon prices. There is little empirical evidence, in a mixed economic system, that public ownership stabilises or lowers land prices. Arguments that it will reduce the price by reducing levels of speculation, allowing cheaper land assembly and enabling lower servicing and carrying costs are unfounded according to Carr and Smith (1975). In any case, demand, and the price of land can go down as well as up. The lesson seems to be that in mixed economies, public sector bodies are often unable to move quickly and clearly enough in conditions of political and economic change to secure their original aims of land purchase. For this reason, many local authorities in Britain have found themselves holding an embarrassing surplus of land, often vacant or derelict, which they acquired at high prices during the land and property boom of the early 1970s.

This brief summary of the arguments for and against public land ownership suggests immediately that the main issues are ideological rather than technical. On the one side are those who advocate public ownership of land for broadly political and social reasons connected with notions of power, collectivisation and equity, and on the other side are those who defend private property, individual rights and the operation of the free market.

Different ideologies on the role and function of the state and its relationship with capital provide different rationales for public land ownership. Hallett (1979) suggested that much of the discussion of urban problems by community groups and journalists and some academics uses implicitly Marxist concepts. This leads to the suggestion that capitalist, land-owning interests manipulate societal 'wants' so as to achieve high rents, for example in the CBD. There is a body of writing which sees the state in this sphere as being principally concerned with the maintenance of conditions favourable for capitalist production and accumulation (Castells 1977; Dear and Scott 1981; Saunders 1981). In the pursuance of these aims the state is led to acquire land, particularly for the provision of basic infrastructure, housing and social facilities. These services are then charged to all capital units via taxation.

The debate about taking land into public ownership does not exist at a theoretical level only. In Britain the past forty years have seen the political pendulum swing to and fro as successive governments have attempted to translate their ideologies into policies dealing with land ownership and development. Interest in land ownership has been central to the debates about local/central government relations, state intervention and the privatisation of certain services. The ability of different levels of government to achieve certain ends through the medium of the public ownership of land has become a key issue in our understanding of the development of cities.

In Britain there have been three major, but short lived, attempts to establish comprehensive public ownership of development land, together with numerous other pieces of legislation for more specialised situations. The first large scale attempt, following the recommendations of the Uthwatt Committee (on Compensation and Betterment), was embodied in the Town and Country Planning Act of 1947, which effectively took all development rights and values into public ownership. Under this act, land was not actually nationalised (although local authorities were given enhanced powers of compulsory purchase), and land owners were free to decide whether to sell their land or to retain it in its existing use. However, the right to develop land was taken from land owners and vested in the state through a system of local authority planning permissions. In exchange for the loss of devlopment rights, land owners were entitled to limited compensation from a Central Land Board. Where development permission was granted, the land owner was required to pay a betterment levy, or development tax, which was set at 100 per cent. In the short term, private development virtually ceased and the scheme became unworkable. It was abandoned in 1953.

A second attempt followed in 1967, with the establishment of a Crown Land Commission which was to take into public ownership, by agreement or compulsory purchase, any land needed for development, at existing use value. The Land Commission would then either develop the land itself, or sell it on at full development price. Like the Central Land Board, the Land Commission was created as a central government body, largely because local authorities were thought to lack the necessary entrepreneurial skills. In the event, the supply of land fell, prices rose, and even local authorities withheld land from the market. The scheme was even shorter lived than its predecessor, being abolished in 1971 after having bought just 1,538 ha of land and sold only 324. In the view of Cox (1984), this scheme failed because the Labour government lacked the commitment to follow it through, because it alienated local authorities who thought that central government was encroaching on their responsibilities, and because it upset builders and land owners who saw it as the beginning of nationalisation.

The period which followed saw massive increases in land prices and intense speculation. These were largely responsible for the institution of a third scheme of public land ownership, the Community Land Act (CLA) of 1975. This empowered local authorities in England and Wales to acquire and develop land, with the ultimate aim of requiring them to consider the purchase of all development land at current use value. This would then be made available to private developers at market value on either a freehold or a leasehold basis, and, in theory, the whole process would have been carefully co-ordinated with the local authorities' traditional land use planning role. Associated with the CLA was Development Land Tax, designed to tackle the betterment issue. Between April 1976 and April 1978, the CLA prompted the purchase of 924 ha of land in the whole of England, and the disposal of just 69 ha (Sant 1980). By the end of 1977 the scheme had lost credibility and it was abolished in 1980, although a very similar arrangement was left intact in the form of the Land Authority for Wales.

Each of these pieces of legislation can be seen very much as products of their time, especially the 1947 Act which owed much to the residual wartime feelings of consensus and centralised planning. The 1975 Act can be seen largely as a direct result of the excesses of the early 1970s land and property boom. As well as having practical planning advantages and disadvantages, each of the proposals had political and ideological overtones; each was introduced by a Labour administration and repealed by a Conservative one. That all three failed to survive can be put down to a mixture of a lack of funds, a heavy and inflexible bureaucracy, some lack of co-operation between central and local bodies and, most significantly, an inability to ensure a steady supply of land or to promote development.

All three programmes attracted intense controversy and produced coalitions of interests for and against public ownership. Those in favour included broadly the political left, most planners, managers of nationalised industries, the trades unions and some local authorities. Those against included the political right, land owners, builders/developers, financial interests, most private businessmen and some of the growing population of home owners. The failure of all three attempts effectively to take development land into public ownership, suggests that the idea may not have widespread political support, but it also reveals something about the relative strengths of the interest groups involved.

After 1979 a new era of reduced public sector intervention took place in a number of countries. In Britain, the incoming Conservative government identified the public sector as an undue burden upon the wealth producers and it determined to return many state run services to the market. One of the first steps was the Local Government, Land and Planning Act of 1980. This gave land policy a firm push away from public ownership by its encouragement of the sale of council houses and the requirement for local authorities to establish registers to promote the sale of surplus public sector land. Subsequently, the general policy of demoting public sector activity in favour of the private sector has been strengthened in certain areas by the creation of Enterprise Zones and Urban Development Corporations. Paradoxically, in the latter case, public sector bodies are used to assemble and deal with land for development in very similar ways to those under the discredited Community Land Act.

There is no doubt that the past decade has seen, in many countries, a very strong and broadly based trend against public ownership. Whilst this trend may be checked or weakened by many circumstances, there is little indication that political electorates have much enthusiasm for old fashioned, monolithic state ownership. Even so, from time to time variations on the theme of municipal ownership of land are raised for discussion. For example Balchin and Bull (1987) suggested a modified CLA scheme with abandoned use values being the basis of public acquisition, or pre-emption zones in which private owners have to give the local authority first refusal on land at current use value. These might operate in a similar way to the *zones d'amenagement differe* (ZAD) and other public land acquisition measures currently existing in France. A fresh attempt to institute a CLA has been argued by Allinson (1988) and in 1986 Tony Benn, without success, proposed a Common Ownership of Land Bill.

Throughout the ebb and flow of these political programmes, there have been, and remain, many other more restricted schemes to deal with specific situations. In the 1930s, for example, land was taken into public ownership for the creation of London's green belt, and after the war large areas were acquired for the new town programme. Currently, many local authorities and other public bodies in Britain are involved in purchasing and developing land in a variety of urban renewal and partnership schemes, and derelict land clearance programmes. In the majority of these cases, however, the local authority land measures can be seen as complementing, not competing with, private development (Needham 1983); in effect, they are supporting the market.

Patterns of public ownership

In a small number of nations, leaving aside the Eastern Bloc, public ownership of land is high. In Israel, for example, approximately 90 per cent of pre-1967 land is publicly owned and development is possible only through leaseholds. Similar arrangements, albeit at lower levels, apply in Hong Kong, Singapore, and, for historic reasons, in parts of the state of Maryland. Generally, however, the public ownership of land in urban North America is very minor; there is no strongly felt need, it would now be prohibitively expensive and it would conflict with the ethos of freedom and independence.

Amongst those areas where municipal ownership of land is favoured, it is Sweden which has one of the longest histories. Since 1880, the city of Stockholm has pursued a programme of municipal land purchase. Today it owns 74 per cent of the land in the city, including practically all undeveloped plots, together with some 50,000 ha outside of the city. Since 1950 municipal land ownership has been stressed as a virtue in itself. Other towns, including Malmo, have followed (Figure 5.2) and by the mid 1970s, approximately four-fifths of all housing being built in Sweden was on land bought or leased

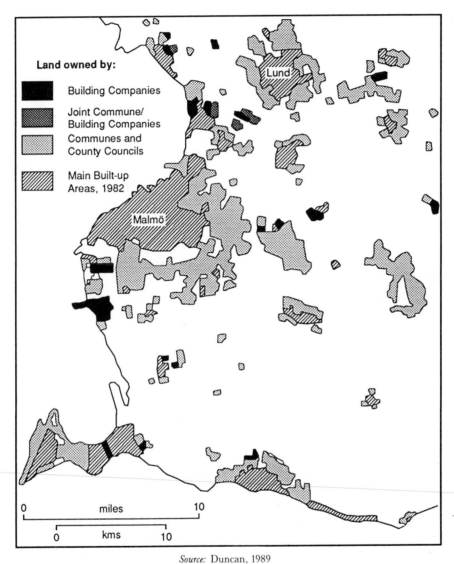

Land owned by:

Building Companies

Joint Commune/
Building Companies

Communes and
County Councils

Main Built-up
Areas, 1982

Lund

Malmö

0   miles   10

0   kms   10

Source: Duncan, 1989

Figure 5.2 Land ownership in the Malmo metropolitan region, 1982

from municipalities. Leaseholds are strong, but land rents are reviewed every
ten years. Stockholm is a fairly small and slow growing city and in some
senses municipal land ownership has worked well. It has planning advantages (Duncan 1989) and the city has gained a good return from increasing
land values. However, it is worth noting that only 10 per cent of the city's
dwellings are single family units, housing is not cheap by European stan-

dards and there is still a shortage of dwellings (Atmer 1987).

Elsewhere in Europe many variants of public ownership exist. In France a number of state agencies become involved in the process of land development, especially in designated development zones around growing cities. In the Netherlands, municipal ownership of land is widespread and land in cities traditionally has been granted on long leases. This is currently cited as one of the reasons for the movement of office development to the edges of major cities where towns such as Amstelveen, Diemen and Hooffdorp near Amsterdam and Cappelle Aan de Ijssel near Rotterdam are willing to sell freeholdland for development rather than insisting on leashold. In some circumstances a tenant can buy off the ground rent for ever and the level of the rent can only be raised if the building density is increased or the land use changes (Blitz *et al.* 1988). There is some pressure for land that yields low rent to have its functions ousted to provide for higher yielding activities. It is tempting to suggest that policies such as these have been responsible for Dutch and Scandinavian cities avoiding the worst problems of urban decay exhibited in Britain and the USA, but in fact there are many differences of historical development and national economic characteristics to be considered. In any case these cities are not immune, as the history of the BANK site between the city centre and the main station in The Hague illustrates. Here a prime site remained grossly underdeveloped and under dispute for two decades before redevelopment started in the mid 1980s.

Outside of Europe few cities have adopted large scale public land ownership programmes. One of the few exceptions is Canberra, where, uniquely for Australia, most of the land is in public hands and is leased for development. The original aims were to keep land costs low, to provide a source of income for public bodies and to provide effective land use planning. The system has a number of oddities (Neutze 1989), and because of Canberra's functions the land ownership is highly fragmented between a number of different government bodies. Many leases are of such long term and with such low ground rents that occupiers act as freeholders. Development, or redevelopment is initiated by the lessees, in contrast to the case in Sweden where the municipality usually takes the lead role.

In Britain the byzantine complexity and individuality of local authority records means that little is known in detail about public land ownership in urban areas. One of the few systemmatic investigations (Dowrick 1974) estimated that in 1972–3, approximately 2.7 million ha of land were in public ownership in Great Britain, representing 11.7 per cent of the total. A broader definition which included roads, common land and leasehold land raised the figure to 18 per cent. Only a very few figures relating to individual towns were available. For example, Dowrick showed that in 1973 the Corporation of Newcastle-upon-Tyne owned 51 per cent of the land in the city, and that the figure for Nottingham was 55 per cent. Bryant (1972) estimated comparable figures for Coventry and Brighton of 33 and 60 per cent respectively.

There exists no comprehensive and accurate survey of public land holdings encompassing such bodies as central government departments, local authorities, state run industries and statutory undertakings. What is known about public land ownership in urban areas is thus limited and imprecise. Crudely, it could be condensed into three broad statements, viz:

1   In large urban areas the majority of land is in public ownership, with the local authorities alone commonly owning more than half of the total.
2   Many public bodies have become involved in land ownership for a wide variety of reasons, and the resulting pattern of ownership is very fragmented.
3   The public ownership of urban land increased rapidly in the postwar years, most especially between 1959 and 1975, in line with the increasing scope of central and local government functions, but it started to fall in the 1980s.

These statements are largely confirmed by one of the few detailed case studies to have been undertaken, that of Manchester (Kivell and McKay 1988). Since 1815, Manchester City Council, and its predecessors, have been actively involved in the acquisition, and to a much lesser extent the disposal, of land in order to perform an increasing range of services. As a result, the most distinctive morphological elements of the city, including the town hall complex in the centre, major areas of parkland and extensive local authority housing estates are easily recognisable as being publicly owned. The amount of land owned freehold by the City of Manchester at 1 July 1982 totalled 8,458 ha. Approximately one-fifth of this land lay outside of the city boundary, but the balance, 6,762 ha represented 58 per cent of all land within the city. Major acquisitions began in the mid nineteenth century in the form of land for parks, but the housing and sanitation needs of the city soon became paramount. By the turn of the century the city owned 1,400 ha – Figure 5.3.

During the interwar years the pattern of land purchases changed and the total purchases grew rapidly. At this time the population of Manchester peaked and this, in combination with slum clearance schemes, increased finance for public housing and the growing suburban aspirations of the residents led to a phase of massive outward urban expansion. Principally, this took the form of large scale, low density council housing estates which involved the City in purchasing large tracts of land in (then) peripheral locations. In the postwar period Manchester's total land holdings continued to grow (Figure 5.3). Two distinct phases can be recognised. First, pre-1970 when purchases far exceeded disposals and resulted in an increasing net total, and post-1970 when rising levels of disposals and falling purchases led to the first ever falls in the net total. The first period coincided with the large scale local authority housing programmes undertaken by practically all British cities, and the second with a regime of tighter financial controls and a transfer of certain activities away from local authorities.

117

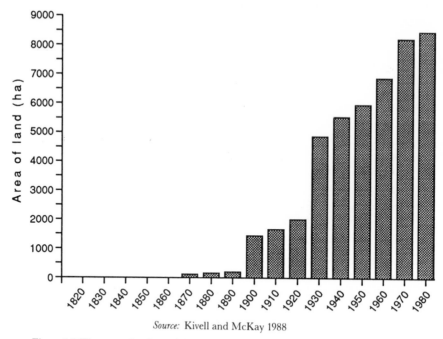

Source: Kivell and McKay 1988

Figure 5.3 The growth of municipal land ownership in Manchester, 1820–1980

The geographical pattern of land owned by Manchester City Council is indicated in Figure 5.4. The pattern is an uneven one with a strong bias towards the southern half of the city. The major determinant is the authority's housing programme and it is the Housing Committee which is the council's largest landholder, being responsible for over one-third of the total (Table 5.6). The city council accounted for almost 90 per cent of the publicly held land in the city, but the balance was the responsibility of more than a dozen other bodies (Table 5.7). The largest of these was the British Railways Board with a concentration of land within 5 km of the CBD, reflecting the way in which the Victorian core of the city became surrounded by railway stations and yards. The Regional Health Authority came next in order, and its total of 160 ha is readily explained by the needs of thirteen hospitals plus a number of clinics, offices and residential homes.

The amount of freehold land owned by public sector bodies in Manchester in 1982 (Table 5.7) totalled 7,635 ha, representing 65.4 per cent of the area of the city. The figures are certainly conservative in that they do not include the majority of roads nor do they cover land held by leasehold and other tenancy arrangements. The overall pattern of publicly owned land is dominated by housing schemes, notably those built in the 1920s and 1930s and again in the 1950s and 1960s, while more localised concentrations can be explained by the presence of public utilities, or specific activities such as the

Table 5.6 Land holdings controlled and administered by Manchester City Council committees, 1982

Committee	Area (ha)	% of total city council land
Housing	2,862.4	33.8
Land and development	2,335.7	27.6
Recreation	1,097.1	13.0
Cleansing	1,080.7	12.8
Education	508.4	6.0
Joint Airport Authority	403.0	4.8
Other	171.0	2.0
Total	8,458.3[a]	100.0

Source: Kivell and McKay 1988
Note: [a]Includes 1,696 ha outside of city boundary.

massive higher education precinct or the international airport. Notable gaps in the pattern of public ownership occur in the CBD where the majority of land remains in private sector ownership. In general terms, the public sector is a very large land owner in Manchester, holding approximately two-thirds of the total land.

Although few cases have been comprehensively documented, it is clear that urban local authorities in Britain are deeply involved in their local land markets. Oxford, for example, purchased large areas in the 1950s, mainly for their own development schemes. Some was used for this purpose, both in the inner area and at Cowley, but some sites were sold in the 1960s for private development, only to be bought back by the city council under a different administration in the 1970s. Sheffield also became widely involved in land purchases, especially for housing, but also for city centre redevelopment. Indeed, in the 1970s the city became a serious speculator in the office boom. It also bought, assembled and serviced land to the south east of the city for industrial development, but was unable to follow through and had to sell it to private developers (Montgomery 1987).

The political and financial climate of the 1980s resulted in attention on public land being switched from acquisition to disposal, although one or two public bodies had been shedding surplus land for some time. Between 1964 and 1979 British Rail sold 32,000 ha while the National Health Service disposed of 730 ha between 1975 and 1979 (Bailey, 1987). From 1980 onwards, government pressure on public bodies mounted, in some cases Whitehall issued direct orders requiring them to dispose of land. Between 1986 and 1988, fifty-two such orders, totalling 110 ha, were issued (Hansard 1988).

Table 5.7 Land located within the City of Manchester owned (freehold) by public bodies, 1982

Public body	Area of land (ha)	% of total city area	% of total landholding
Manchester City Council	6,762.3	57.9	88.6
British Railways Board	337.5	2.9	4.4
NW Regional Health Authority	160.3	1.4	2.1
Greater Manchester Council	134.2	1.1	1.8
University of Manchester	74.6	0.6	1.0
UMIST	43.6	0.4	0.6
British Waterways Board	35.4	0.3	0.5
NW Gas Board	26.0	0.2	0.3
Gtr Manchester Passenger Transport Executive	18.6	0.2	0.2
Central Electricity Generating Board	15.6	0.1	0.2
British Telecom	15.4	0.1	0.2
NW Postal Board	7.0	–	0.1
NW Electricity Board	3.3	–	–
BBC	1.5	–	–
Total	7,635.3	65.2	100.0

Source: Kivell and McKay 1988

Public/private partnerships

In the process of urban development and redevelopment a variety of partnerships between public and private sector interests have become increasingly common. A number of European countries have long favoured such arrangements, but they have been particularly encouraged recently in both Britain and the USA as a way of overcoming some of the most difficult land planning issues involved in urban redevelopment. In such cases the actual ownership of the land is usually quite clearly in one sector or the other, but the involvement of both sectors in joint programmes of development blurs some of the traditional boundaries. Many variants of partnership exist, but the commonest pattern is for the public sector, in the form of a local authority or a government body such as a development corporation, to acquire adjacent plots of land in order to assemble a large site, to reclaim and service it where necessary, close streets and attend to transport matters and to provide a (perhaps 'streamlined') planning service and financial incentives such as loan guarantees and tax concessions. The private sector partner typically provides the bulk of the finance, the design and technical skills, undertakes the actual building and construction work and takes the lead in marketing the development. The public sector may retain a long term interest, for example in the form of the freehold.

Given that the public sector has often been blamed for delaying development, for example through excessive bureaucracy or an unwillingness to

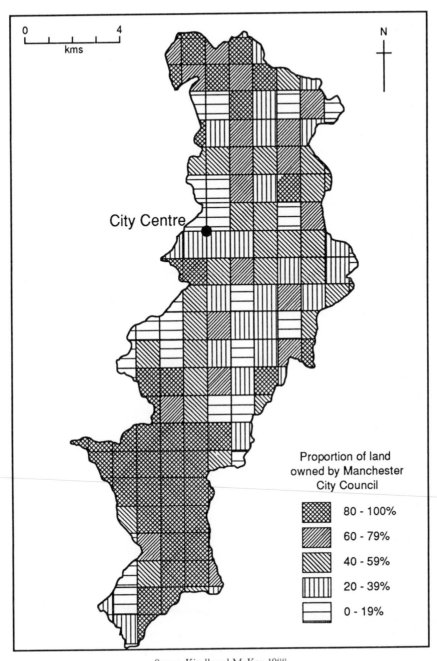

Source: Kivell and McKay 1988

Figure 5.4 Land owned by Manchester City Council in 1982

dispose of its own land, this kind of a partnership can be seen as an effective way of harnessing complementary resources and speeding development. Many of the most widely publicised redevelopment schemes of recent years have involved partnerships. There are, however, also some dangers in the partnership arrangement. Above all concern has been expressed about the accountability of such public sector bodies as development corporations which may be making very large investments of public funds and which will profoundly change the structure of local communities, yet they do not have the traditional accountability through the electoral process which local authorities must face. Second, there is the risk that the public bodies are merely guaranteeing business profits and providing an attractive local economic climate for the private sector participants. There is the possibility of a major conflict between capital and community interests, especially if the local authority role as development partner becomes confused with its role as planning authority. Further dangers exist in the possibility that, if development permission is refused or if concessionary conditions are removed at a future date, the private developers may simply pull out and move their capital elsewhere, whereas public bodies are immobile (Cummings *et al.* 1988). Finally, there is asymmetry of information: public officials have to be open and are inclined to grant concessions up to the statutory limit, whereas private developers can deploy a more secretive business strategy.

Whilst these criticisms clearly have weight, there is no doubt that partnerships have attracted large sums of private sector investment and activity into areas where, in many cases, land had lain idle and neglected for many years. The accountability of the public sector is not always obvious, but neither is it entirely lacking. Often there is local authority involvement, with eventual electoral accountability, but even development corporations are answerable to central government and perform in similar ways to the boards of nationalised industries or new town development corporations.

CONCLUSION

The conclusion to this relatively lengthy discussion of ownership can be quite brief. The notion of ownership when applied to land is a complicated one, involving not just the historical intricacies of different forms of legal tenure, but also different packages of rights to occupy, use and dispose of the land. Ownership of land is important in shaping many aspects of urban development, but it is also an integral and deeply entrenched part of the economic and political organisation of different societies. Indeed the pattern of land ownership is one of the key diagnostic variables in distinguishing between different socio-political systems. In many respects, the political spectrum from free market to centrally planned economies is exactly shadowed by the spectrum from private ownership to state ownership of land. Not surprisingly, therefore, in the mixed economies which dominate the western world, the ownership of land is split between public and private sector interests, with the respective shares varying according to local circumstances. As the scale and complexity of central and local government functions grew in the 1960s and 1970s, the amount of land in public hands also increased. More recently, however, especially in Britain, the public sector landed estate has been cut back significantly. Almost everywhere in the cities of the developed world, it is the financial sector, the banks, insurance companies and pension funds which have accounted for the largest growth in land and property ownership, and it is their behaviour which is playing an increasingly important role in shaping the city. It is also, to a large extent, the behaviour of land owners, the way in which they accumulate social and economic power and the balance of gain and loss between them and the community which shapes the land policies in different countries. It is to this question of land policy that the next chapter will turn.

6

LAND POLICY

Land policy is a wide set of activities whereby governments seek to influence the use, planning, ownership, price and benefits of land, especially within the process of development. Central to this is the question of allocating the enhanced value which accrues from development. A comprehensive overview of land policy will not be provided here for it is available in a number of well established texts (see, for example: Ratcliffe 1976; Darin-Drabkin 1977; Lichfield and Darin-Drabkin 1980; Barrett and Healey 1985; Hallett 1988).

The purpose of this chapter is more limited. It is to review briefly the process of land development and then to examine some of the main principles and alternatives within urban land policy. Some attention will also be given to the way in which such policies have become more fragmented and *ad hoc* in recent years, and examples will be given of the application of policies in Europe and the USA. The discussion will place land policy firmly within a geographical context, because of the geographer's central concern with spatial patterns and relationships. At the same time it will show that the focus of the land policy debate has shifted somewhat over the past decade, away from the 'instrumental' and narrowly land use approach towards a broader political economy approach as discussed by Cox (1984), Barrett and Healey (1985), Ambrose (1986), Rydin (1986) and Healey *et al.* (1988).

The normal purpose of land policy is to control development, either in the sense of shaping land use patterns, or in the broader sense of ensuring a degree of fairness and redistribution of the gains to be made. Traditionally the emphasis has been upon guiding or restricting new development and it is clear to see, in the USA for example, that there has been considerable pressure for new policies in areas of rapid urban growth. In this sense, land policy evolves as a response to issues raised by the process of development, and it may originate from narrow land use concerns, or from broader fiscal, social or ideological considerations. In the past two decades, however, it is not solely *ab initio* development which has been the focus of land policy, but also the need to redevelop derelict and decaying parts of the city. Since development, in one form or another, is clearly at the heart of the land policy debate, it will first be necessary to outline something of the development process itself.

124

THE DEVELOPMENT PROCESS

The process of development varies greatly, especially according to the level of public intervention and the political economy in which it is set, but generally it follows a predictable sequence. This starts when urban/economic growth stimulates the need to develop land to a more intensive use. It is followed by an expression of interest by developers (sometimes including the state), the preparation of proposals and plans, possible changes in the ownership of the land, the securing of finance, physical preparation of the site and construction work and finally the occupation of the completed scheme, either by the developer or another owner. This process remains essentially that summarised by Lichfield (1956) and Drewett (1973) for greenfield sites. Barrett *et al.* (1978) simplified it into three stages: 1) development pressures and prospects; 2) development feasibility; and 3) implementation. For urban redevelopment, Bourne (1967) provided a succinct review, but much has changed since then. In particular, for the inner areas of many of the largest cities in Britain and the USA, the demand to develop or redevelop land during the 1970s and 1980s was either very low or non-existent. Not only had the restructuring of the city resulted in a lowering of demand, but in many cases a high level of public ownership had effectively stultified the market, and the existence of rigid use classes and land use rights in perpetuity had fixed an unrealistic floor price for the land (Chisholm and Kivell 1987).

The duration of the development process may vary from a few months to decades, and this itself is partly determined by the ruling land policy. The

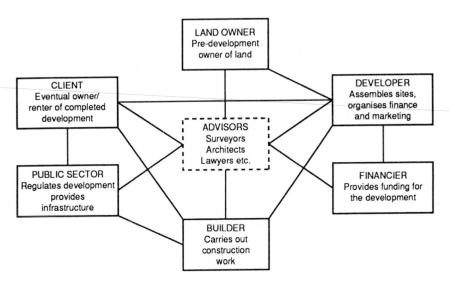

Figure 6.1 Relationship between principal participants in market governed land development

125

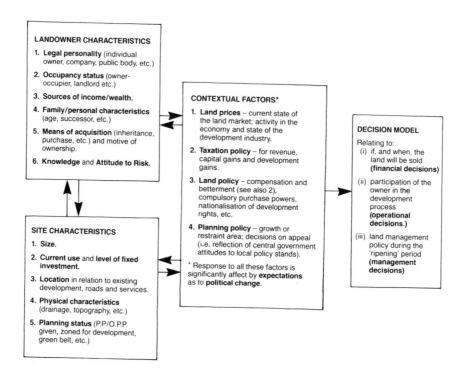

Source: Goodchild and Munton 1985: Figure 1.2

Figure 6.2 Land owner behaviour: constraints and the development process

main participants are the original land owner, developers, professionals such as surveyors and architects, builders, financiers, central and local government officials, and the eventual occupier of the development (Figure 6.1). The roles of these participants may ebb and flow during the course of the development, there will be considerable overlap and each of the sets of actors may play roles in addition to their principle one. A clear analysis of these various elements is that provided by Goodchild and Munton (1985) who placed particular emphasis upon the land owner. Figure 6.2, adapted from that source, gives a concise view of the conditions influencing land owners during development. Since it is the land owners and developers who commonly form the focus of land policy, their roles warrant further discussion.

In the USA, and other countries with relatively weak land policies, developers and speculative land owners may begin to show interest in land on the urban fringe long before it is actually needed for urban development, and there may be many transactions in land with future 'hope value'. In those European nations with a strong tradition of public ownership of urban land, for example Sweden and The Netherlands, such behaviour by private owners is pointless because the municipalities have effective pre-emption rights. In Britain, as so often, there is a compromise between these two positions. Speculative behaviour on the fringe is minimised by planning guidelines which result in there being two, almost separate, land markets, one in agricultural land with virtually no possibility of development, and the other in sites which have, or are likely to have, development permission. The relative firmness of land policy measures such as development control and the green belt removes much of the speculative behaviour.

The developer is a key actor because, in most instances, this is who initiates the conversion of a site, puts the deal together and oversees it from start to finish. Developers often specialise in particular activities, for example housing, office development or shopping centres and this, together with strong planning guidelines, means that different kinds of land sub-markets with different spatial patterns commonly exist. The developer's key decision is to purchase land and push it through the development process. Depending upon the size and type of the project, this may happen on an individual site-by-site basis, or the developer may acquire and hold adjacent plots until a large site has been assembled for comprehensive development, or he may establish a land bank consisting of many sites to ensure a steady flow of future activity.

The timing of this development will be determined both by the state of the national economy and the level of local demand, and fluctuations in these two factors have been seen clearly during the past decade. It is local demand which will largely determine the overall financial equation, especially the price which the developer will be willing to bid for the land. In most cases, this will be arrived at through a residual calculation along the following lines.

$$L = S - [l + c + i + e + p]$$

Where

L = developer's bid price for the land
S = anticipated selling price of completed development
l = estimated land preparation and infrastructure costs
c = estimated construction costs
i = estimated interest charges on land and construction costs
e = estimated legal and marketing costs
p = estimated developer's profit

The anticipated selling price, S, will be largely independent of the costs within the brackets, being determined primarily by the market value of existing properties in the local area. Only if the sum of all of the estimated costs, including those of acquiring the land, falls below S will the development take place.

Most commonly the developer is a private sector individual or corporation, but across Western Europe and North America a wide range of private and public sector combinations may be observed. The developer may be purely private sector, working within public sector guidelines which range from relatively lax (e.g. USA and Australia) to relatively tight (e.g. UK and West Germany). Alternatively, the public sector may be involved at an early stage in the purchase, assembly and servicing of land to be sold to private developers. This is the normal state of affairs in such countries as Sweden and, to a lesser extent in France, and it has been used increasingly in the USA and UK to deal with problematic inner city redevelopment. The public sector may also go considerably further than this, for example, by retaining a partnership interest with the developer, or by doing everything itself from initial land purchase, through the development stage to ownership of the completed project. This latter strategy can be seen in the extensive public sector housing schemes in many European cities.

In the case of private sector schemes, the original land owner is normally different from the developer, although as a project proceeds, ownership commonly passes into, and then later out of, the hands of the developer. Whereas the developer's key decision is that of buying land, the land owner's crucial role is normally that of selling his land, especially with regard to such questions as when, how much, for what price and to whom. Again Goodchild and Munton (1985) have provided an excellent analysis.

Most owners of undeveloped land, unlike the planners, builders, estate agents and developers who are involved, are not professionals. They are not concerned with land dealing as a business. They may be farmers, householders, industrialists with surplus space or individuals who have inherited small areas of family land. Even public authorities are sometimes complete amateurs when it comes to owning, managing or selling land. The result of this is that a wide variety of unpredictable, sometimes irrational, even perverse, behaviours take place. Table 6.1 summarises reasons for selling land for development, based upon British experience. Personal factors are shown amongst the reasons for selling, but considerations such as sentiment, conservatism or family connections may also delay such sales. These influences are particularly strong amongst farmers, few of whom systematically evaluate their financial return from agriculture against alternatives. Farming is a way of life as well as a business and farmers tend to take a long term view. Nor should this be dismissed as mere sentiment, for most farmers have a deep seated concern for the land which prevents them from seeing it, as a developer might, simply as a commodity.

To some extent the relationship between development and land policy is a circular one. The behaviour of land owners and developers influences the derivation of policy, but it is the kind of policy which is in force that partly determines their behaviour. In the USA the general strategy of land owners differs from that in the UK, largely because planning and fiscal policies differ. For example, in much of the USA land is taxed recurrently on its

Table 6.1 Reasons for selling development land

Category	Elaboration
1 Financial gain	Sale determined by financial factors alone, e.g. because the price is too good to refuse, or because there is a more attractive investment for the owner's capital.
2 Need for cash	A sale which the owner is forced into because of a pressing need for cash.
3 Property obsolescence	A sale to move out of premises which are old and/or expensive to maintain, or which are no longer suitable for their present use due to technological changes, or have become unsuitable from a change in the neighbourhood rather than the premises themselves.
4 Personal	A sale for non-pecuniary reasons which are personal to the owner, e.g. a wish to retire or live in a different location, or a move forced by illness, or often, following the owner's death.

Source: Goodchild and Munton 1985

capital value, taxation thus rises with increasing values as development prospects come nearer and the owner will thus be encouraged to develop. In Britain, the absence of a direct holding cost means that owners are rarely encouraged, or forced, to sell because they can no longer afford to retain ownership. In Britain the key event is the granting of planning permission. Since this is when the real value of the land is normally maximised, it should, logically, be the event which prompts the sale or development of the land. Within the urban area the markets and the stimuli for redevelopment are more complex, because in addition to the overall state of the economy or the local rate of urban growth, there are factors such as the differential shift in demand for various uses and the question of locational or property obsolescence to be considered.

REASONS FOR LAND POLICY

All governments intervene in the land market, and in the process of development, although to varying degrees. Generally, the justification for this is twofold, being based upon the belief that it reduces inefficiency and improves equity. The argument that the free market would allocate land to its most desirable use without intervention is true only under conditions of a perfect market and entirely equitable income distribution. In reality, neither condition obtains and governments feel obliged to intervene in the form of a land policy. According to Hallett (1988), whatever the ideological arguments, the practical reality is that laissez-faire urban development does not produce satisfactory results, and some degree of intervention becomes necessary. An alternative view is that in general public intervention corrects some instances of market failure, fails to correct others and creates yet others (Lee 1981). It is also important to note, however, that the use of the term land policy implies a coherence and clarity which does not exist in reality. Until recently it was suggested that land policy had no clear and accepted meaning (RTPI 1985) and that what most countries possess is a set of policies which are evolutionary, partial and ad hoc.

At a general level land policy is easy enough to define as the framework which relates land ownership, land values, land use planning and development policies, but at an operational level there is much variation. Barrett and Healey (1985) suggested that land policy should be seen, not simply as a set of regulations for achieving narrow land use planning objectives, but as an integral part of broader economic and social policies, although of course these in turn may be imprecisely specified. For this reason it is important that land policy should not be determined solely by economists, administrators, surveyors and other land professionals, but that it should also have an input from geographers. Again, this should be seen not narrowly in locational terms, but as a part of the broader patterns and relationships involving land within society.

The distinction between equity and efficiency in the derivation of land policy remains relevant, but an equally important distinction may be made between aspects of control and promotion of development. Control involves the regulation and limitation of the free market and some of the powerful agents within it. Promotion involves the encouragement of desirable forms of development. It is possible to argue that, in the past, control has had some perverse side effects, such as the separation of home and workplace, rising property and land prices and the exclusion of low income groups from much of the suburban housing market (Hall 1974). In the early 1980s, however, a slump in development of almost all kinds resulted in the pendulum of land policy swinging towards the promotion of development, especially as a means of facilitating economic growth. In the UK a pro-development ethos was promoted and the government's avowed stance was 'to keep to a minimum any involvement in the operation of the land market' (DoE 1988: 33).

Aspects of control

1) *Externalities*. Land policy needs to be able to regulate externalities which lead to a loss of welfare by the general public, or by individual third parties, caused by development where only the private costs and benefits have been accounted. At a small scale this may involve localised aspects of traffic or building regulations, but at a larger scale there are also developments which could be undesirable neighbours, for example an airport, a high speed rail link, a power station or a refuse tip.

130

2) *Powerful agents*. An undue concentration of landed power in the hands of a few individuals or corporations is potentially dangerous in both economic and social terms. Problems of monopoly land ownership have been discussed in the previous chapter, and Ambrose and Colenutt (1975) have written persuasively about the concentrated power of developers. It is but a small step to extend the argument to the 'manipulated city' hypothesis whereby powerful capitalist interests can work to the detriment of weaker social groups and desirable land use activities, for example by commercial development squeezing housing out from urban central areas. These arguments have been used in a number of recent dockland redevelopment schemes, including those in London and in north east England. Cox (1984) argued that working class groups are weak and poorly organised in the UK and have little influence on land policy. He rejected the view of Massey and Catalano (1978) that the Community Land Act was the result of a groundswell of community action groups and trade union activity. Further evidence supporting the need to curb the role of powerful agents was provided by Harrison (1983) who claimed that land speculation provoked the 1974 recession through an unbalanced flow of land onto the market, the distortion of production costs of firms and the reduction in the spending power of households. An alternative view which has gained some currency more recently is that the UK economy has been unbalanced by an import based consumer boom sparked off by people taking financial equity out of their inflated property values.

3) *Urban sprawl*. The control of urban sprawl has been one of the main planning priorities of a number of governments, including that of the UK (DoE 1988: 33). It is also an aspect of land policy which puts it into the mainstream of geographical analysis. The operation of this process has been comprehensively documented by Hall *et al.* (1973). Even in North America, where attitudes towards urban sprawl have traditionally been fairly relaxed, a number of municipalities have recently passed 'no-growth' or 'low-growth' land policies. In Britain and the rest of Europe, such policies remain important, although the justification for them has shifted from the protection of agricultural land to the safeguarding of environmental quality.

4) *Prices, profits and gains*. Sudden, or large, increases in the price of development land, especially where they can be linked to housing shortages or price inflation, commonly bring forth demands for tighter land policies, even though it is far from clear whether high land prices cause high house prices or vice versa. There are two interconnected aspects to this complex issue of prices, profits and gains. The first is the extent to which the state should compensate land owners where it restricts their right to develop their own land, or where it takes land for public purposes. The second is the principle of whether, and the level at which, the state should take a share, through taxation or otherwise, of increases in land values, however those increases are caused. For Hallett (1988), one of the essential criteria to be applied to land

policies is the question of whether they provide a reasonable level of taxation of the gains accruing from land ownership.

Aspects of promotion

Land policy measures are sometimes assumed to be mainly concerned with controlling undesirable aspects of the free market, but a balanced view requires that they should also be addressed to the positive promotion of desirable development.

1) *Development and redevelopment*. According to the Royal Town Planning Institute (1985), land policy in Britain has been increasingly concerned with enabling and assisting private sector development to take place. The central question by which such policy has been judged is 'How can we get more land into development?', and this has guided urban renewal and redevelopment policies even more than those relating to new development. Again, one of Hallett's key criteria was 'Do they facilitate an adequate supply of building land for housing in aggregate, and access to an acceptable level of housing for the poor?' (1988: 196). In their desire to promote development, governments have been inclined to grant various incentives and concessions to developers, especially in the form of 'pump-priming' investment and in the relaxation of planning rules and bureaucratic delays where these have been seen as obstacles to development.

2) *Good town planning*. Recent attempts to promote development in order to stimulate local economies may be open to criticisms of opportunism, and a number of them have failed to sustain their early promise. It is particularly important therefore that the role of good planning practice, Hallett's (1988) criterion of 'liveability', should not be overlooked. Exactly what makes for good planning practice is debatable, but it must include an understanding of both the need to defuse the potential for social and political unrest in urban slums, and the quite genuine and altruistic motives of a series of middle class reformers from the late nineteenth century to the present day. Much of that individual reforming zeal has today become institutionalised, finding expression in the voices of professional organisations and local authority committees. The worst of the urban slums, and their ubiquity in western cities, have now been removed, so the focus of the urban reform movement has shifted to a concern with traffic problems, a human scale for the city and the nature of its physical and social environment.

3) *Public goods*. Some of the earliest land policies, albeit very much on an ad hoc basis, were concerned with compensating owners for land taken by the state for public or community purposes such as roads, public health or defence. Today a number of urban land using activities fall, more or less, into this category of public goods, most notably those connected with the provision of transport infrastructure.

4) *Redistribution and welfare*. One of the broad concerns of land policy, and

132

indeed one of the starting points in its establishment, has traditionally been a desire to achieve some measure of equity in the distribution of resources and opportunities. The need for this kind of policy stems from genuine humanitarian concerns by governments to protect weaker groups within society but also from the needs of governments to protect their own stability and legitimacy. It is also true of course that land transactions can provide a convenient focus for taxation and revenue raising measures. Some land policies with these aims in mind are couched in very general terms, aimed, for example, at recouping for the community some of the financial gains from land development. Others have more specific purposes, such as encouraging the spread of home ownership.

INSTRUMENTS AND TECHNIQUES OF LAND POLICY

The instruments and techniques of land policy may also be reviewed under the headings of control and promotion of development. The relative emphasis upon the two divisions varies over both time and space, but it is fair to say that whereas techniques of control have traditionally been the most powerful aspects of land policy, and still dominate much of the academic discussion, it is techniques of promotion which have been gaining currency in government circles, not least in Britain. Some preliminary discussion of land use planning has taken place in Chapter 2, but it is necessary to return to the theme here in order to explore the wider context of land policy.

Control

Land use planning and zoning

The essence of land use planning or zoning, involves the state, usually in the form of local authorities, laying down generalised structure plans, or more detailed local plans indicating where development will be permitted and in what form. The aims are normally to demarcate approved uses, to ensure adequate land for all activities in suitable locations and to avoid incompatible uses. The process is one of regulation and co-ordination, but it is essentially a permissive system which is not able to guarantee that the indicated development will actually take place. This is broadly the form which planning has taken in Britain under the major Planning Acts of 1947, 1971 and 1980.

In addition to signifying the permitted forms of development, such programmes also make outline provision for residential and other land for some years ahead. Associated with the development plan, there is frequently, as in Britain, some requirement for planning permission which can only be given by the local authority. In this way development rights are vested in the state even if the land and the development process remains in private hands. In addition to this blanket planning control, there may be quotas or permits for specific uses, such as the office development permits which were required in London in the early 1970s, or the less formal limitations on the numbers and locations of quasi retail activities which are operated by many planning authorities. At the most detailed level there may also be a set of building regulations to control construction standards; these may operate locally, as in the USA, or nationally, as in Japan.

Since all these planning measures are designed largely to control the free market, much discussion revolves around the success with which they do this. Britain has one of the longest histories of planning, with a set of regulations which were basically framed in the 1930s, but there are many questions about how well they have coped with the greatly changed circumstances of the 1980s and 1990s. Ambrose (1986), for example, argued that had the 1947 planning regulations been as powerful in reality as they appeared on paper, they would have severely limited the rate of capital accumulation through land development. The argument that planning had a restrictive effect upon development was fashionable in government circles in the 1980s, and was used as a reason for simplifying planning in certain circumstances. This in turn gave some credence to the earlier view of Miliband (1973) that decision making processes which were held to be democratic were often strongly influenced by narrow interest groups. In other words, in a capitalist society, even the state has to be primarily capitalist and controls the interests of capital only to a limited extent. Balchin and Bull (1987) have suggested that planning is too much organised around the needs of the property industry and that consequently it fails to achieve either its aims of equity or efficiency. They considered that it restricts the supply of land and pushes up prices by creating a shortage of planning permissions, albeit for the commendable aims of restricting urban growth. Cherry (1991) suggested that although the 1947 system offered the possibility of controlling the urban land use map over a 20-year period, this did not wholly materialise. His view is that, as development has become increasingly opportunity led, planning has lost its claim to be a comprehensive guide to urban form. However, from this common diagnosis that the regulatory planning system in Britain is not effective, the prescriptions were very different; the government argued for less planning, but Ambrose, Balchin and Bull advocated a strengthening of central control.

Finally, it is perhaps worth pointing out, along with Hallett (1988) that possibly too much has been expected of regulatory land use planning. In the sphere of housing, in particular, many problems are not amenable to solution through land policies alone for they stem from a mixture of poverty, distorted tenure mixes and sundry other social issues.

Taxation and other fiscal measures

Fiscal measures may be used in a great variety of ways to guide and regulate

134

the land market and the development process. In the majority of cases the prime aim is to raise revenue, with any effect upon land development being merely a corollary, but a few fiscal measures are designed specifically to influence land use or development.

A simple twofold division into taxes upon property and land *per se*, and those levied upon the development process may be made, but some overlap occurs. Many countries impose a general property tax, either upon the capital value or the imputed rental income of the property. Until recently this was the main way of raising revenue locally in Britain, and with the demise of the Community Charge it appears likely to be applied again. Property taxes have many advantages, including being cheap to administer, difficult to evade (because property is permanent) and serving as a surrogate wealth tax. There is also a weak counter-argument that any direct taxation of buildings deters new construction and maintenance. A rather different form of property tax is that payable by home owners upon the imputed rental value of their own house. Such a system (Schedule A income tax) existed in Britain before 1962, in Germany until 1987 and still exists in The Netherlands.

Taxes upon land, rather than upon property are an alternative approach. For a time in the nineteenth century there was a utopian view which claimed that land tax was the true solution to the land problem and would help to remove many other social ills too. Henry George (1879), in particular, argued that taxes on land should be the sole source of government revenue, and that all net rent should be expropriated in this way but he could not counteract the argument that if all profits are eliminated for the land owner there will be no incentive to develop or improve land anyway. In principle, a site value tax may provide a powerful incentive to persuade owners to use their land intensively, especially if it is based upon an assessed value which takes the local development potential into account. For this reason it might be expected to appeal to land owners in the commercial and productive sectors who already have developed their land to its best use. There are technical difficulties associated with site taxes (Holland 1971), not least in terms of appropriate valuations, and Hallett (1977) suggested that too much has been claimed for them. Nonetheless, it can be argued that their most effective application might be as an instrument specifically targeted on vacant or underused land, rather than as a blanket tax for raising revenue.

The comparative merits and impacts of property and site taxation have not been widely investigated but a few studies in Pittsburgh (Richman 1965), in Australia (Woodruff and Ecker-Racz 1974) and in New Zealand (Clarke 1974) have suggested that there was no marked difference in urban land use patterns under the two systems.

In the UK there are no taxes levied exclusively upon land, but land and property assets are routinely caught within more general systems of taxation. For example, capital gains tax is payable on profits made from land sales, corporation tax and income tax are due on income from rents and dividends, inheritance tax is payable upon the transfer of land at death, stamp duty is payable on transfers and VAT is payable on building repairs, alterations and reconstructions.

In addition to the general land and property taxes outlined above, there are also fiscal instruments which are specific to the process of land development.

Betterment tax, strictly speaking, is levied to cover public investment in improvements which create rising land values, but the term has a more general use as a tax which siphons off some of the profit or 'unearned increment' to be gained from development. Here too there is a complex valuation task especially in assessing precisely what has caused the gain. The question of separating out the specific efforts of the developer from the more general urban processes which have created a situation ripe for development is rarely addressed. The level of the tax is also crucial for if it is punitive, owners will be deterred from selling and the development process will stagnate. From British experience Hallett (1988) suggested that levels of 80–100 per cent will effectively kill the market, but a level of 30 per cent may be workable. Britain had such a development land tax, at variable levels, but it was removed in 1985. The amounts raised were small (£68 million in 1983–4), leading to suggestions that its main aim was political rather than economic. Certainly, it proved to be neither an efficient means of raising tax nor an effective instrument for regulating land development. Development taxes may be levied on realised gains, i.e. when the land is sold, or upon some computation of unrealised gains.

A further small charge may be made in the form of a fee paid for obtaining planning permission or building regulation inspections. Where it is used, this is normally very minor and should be seen as an attempt by local authorities to defray some of their administrative costs rather than a direct tax on development.

A final form of development charge which remains to be mentioned is a rather grey area concerning infrastructure. In some cases, for example in West Germany, a formal charge is levied upon developers to cover public sector infrastructure costs in connection with private development. Where a formal charge is not levied, some element of local bargaining may become common. For example, planning permission for a private sector housing scheme may be easier to obtain if it can be shown to include some public open space or other community facility. The view of the Royal Town Planning Institute (1985) is that the shift towards market determination of land use in the 1980s, and the limitation of local authorities to statutory duties all gave greater scope for bargaining to produce infrastructure or planning gain (called exactions in the USA), paid for by the developer. This process has raised concern in a number of quarters (Keogh 1985).

Land ownership

Taking land into public ownership provides one of the fullest measures of control over development. Many variants are possible, from temporary to permanent ownership, from full nationalisation of land to appropriation of selected development sites and from first moves to last resort. Full nationalisation of land is the most radical policy, but it is not one which is widely used in western countries, partly because of technical problems over leases and tenants rights, but, more particularly, because it would be prohibitively expensive and politically and socially unacceptable in the present climate of opinion. The nearest approaches to this are the Dutch and Swedish systems discussed in the previous chapter. In Britain, blanket policies to take development land into public ownership have been tried and abandoned in the past (see Chapter 5), but some elements remain in selective use.

Promotion

In order to promote the development of land, especially if market forces are not achieving the desired results, there are many techniques which governments can use. The past two decades, which have witnessed the stagnation and restructuring of many urban land markets have also seen the operation of a plethora of policies designed to promote development.

1) *Government as developer.* Government involvement in development can take place at many levels up to and including the use of government bodies to act directly as developers. Central and local governments in western nations are democratically elected, representative bodies with different aims and responsibilities from entrepreneurial corporations, so their direct involvement with the risk-taking business of development tends to be limited. Even so, they do get involved with some very major undertakings. For example, the British new towns programme, conducted almost entirely through public sector agencies, was responsible between 1946 and 1970 for building twenty-eight new towns which today house over two million people. A similar mechanism, in the form of Urban Development Corporations, has been chosen to deal with the most intractable problems of Britain's inner cities. Many other public development bodies exist, for example the Scottish Development Agency, the Land Authority for Wales, Enterprise Zones, English Estates and others, with multiple roles to play in the process of urban development. A further comprehensive example of government as developer occurs in the large scale public housing schemes which are to be seen in many European cities (particularly in Britain), and to a lesser extent in North America. In the majority of these cases the state, normally in the form of local government, takes on all of the developers' roles, including land acquisition, finance, design, planning, construction, ownership and management of the completed scheme.

An increasingly common tendency, in Britain and in other countries, is for the public sector to work together with private companies in a development partnership. In both France and Germany, quasi-public bodies and public–private sector partnerships are active in urban development and renewal, but they are normally answerable to the local authority not to central government.

One form of public land ownership, in the direct pursuit of development, remains to be mentioned. This is the policy of land readjustment, or replotting, commonly seen in cities in Japan and Korea, but relatively rare elsewhere. Under this policy, a public authority temporarily acquires adjacent plots for development from a number of owners. The land is redivided and serviced by the public authority, which retains sufficient plots to cover its own costs and returns the remainder to the original owners in proportion to their original contribution.

2) *Support for the market*. Most governments accept that a completely unregulated free market does not produce the most acceptable overall environment for urban development, and that some degree of intervention is necessary. Measures to stimulate or support the market may thus be taken, either nationally or locally, where there is evidence that the market itself has failed, or where intervention for other purposes may have produced negative effects.

In Britain, attempts to stimulate the market, especially in stagnant inner city areas, have been vigorously applied. The principal way in which this has been done proceeds from the premise that bureaucracy in general, and planning in particular, has imposed barriers to development which need to be lessened. Thus in 1986 a whole tier of government, represented by the Metropolitan County Councils, was abolished in Britain's seven largest urban agglomerations. The government also let it be known, in Circular 22/84 that in future it would be less concerned with the detail of structure and local plans. In selected areas there were concentrated attempts to reduce planning delays and stimulate the market. Simplified Planning Zones were introduced in 1986, with the aim of speeding up development by giving advance planning permission for specific kinds of development in clearly designated areas. The first such zone was established in Derby. Similarly, Enterprise Zones offer simplified planning procedures together with tax incentives and exemption from rates for ten years. Currently twenty-seven Enterprise Zones exist, but most of them will expire between 1991 and 1994 and it is unlikely that they will be extended. A further scheme to stimulate the market and give it confidence is represented by the Urban Development Corporations. Here a comprehensive package of land acquisition and reclamation measures, building refurbishment, infrastructure provision and simplified planning is designed to produce conditions attractive to private investment. In this way a public sector investment of say £X, can be used to stimulate private sector investment of £5X or even £10X. Market stimulation was also the prime motive behind the establishment of Land Registers in 1981 (see Chapter 7).

The above mechanisms are largely concerned with planning matters, but governments also have a range of fiscal policies which may be used, directly or indirectly, to stimulate urban development. Policies relating to taxes and interest rates are powerful means of affecting economic development and both the construction industry and property markets are particularly sensitive to changes in these areas. More specifically, governments may decide to encourage particular forms of development, for example home ownership, through financial support. The mortgage insurance facilities provided in the USA by the Federal Housing Administration and the Veterans Administration, especially between 1945 and 1970, and the tax relief on mortgage interest payments in the UK both contributed hugely to extensive postwar suburban growth.

3) *Grants and other promotional legislation.* Where the local urban economy has collapsed, and the development potential has fallen, perhaps to nil, as in some British and North American inner city areas, larger incentives may be necessary to stimulate the market. Again it is in British cities that the most elaborate, and sometimes confusing, schemes can be found. The 1980s were marked by a plethora of measures which were sometimes complementary but often overlapping, and which had many confusing name changes. The common purpose of these was to improve the development potential of inner city land and to get private developers interested in bringing it back into use. The measures outlined above, including Enterprise Zones and urban development corporations were important, but in 1988 there was also some tidying up of other measures in the form of a new city grant. This replaced several existing grants and aimed to bridge the gap between the costs of a development and its sale value in order to create a socially useful development which would also allow the developer a worthwhile profit.

LAND POLICY: CASE STUDIES

Many variations in national policies have become apparent from the foregoing, but it will be instructive to take a further brief look at contrasting policy frameworks in a number of different countries.

The European Community has no supranational land policy or planning system beyond the individual schemes of its member states (Haigh 1989), but its common agricultural policy and its programme of environmental legislation both provide possible starting points. Specific urban policies are rare, being limited to modest ad hoc finance from the economic and social funds for integrated development programmes. Land use planning is seen as essentially a local activity although federal or regional bodies play important roles in West Germany, Spain, Italy and Belgium.

The United Kingdom

The United Kingdom has a comprehensive, but fragmented, set of land policies. Many of these have been discussed above, and Britain's system is well documented (e.g. Cox 1984; DoE 1988; Healey *et al.* 1988) so only a brief overview to tie things together will be given here. Since 1909 land policy has involved various degrees of state land use planning, conducted by local authorities guided by national legislation, and a mixed but largely private pattern of land ownership and development. State involvement generally increased, especially under Labour administrations, up until about 1980. State intervention was however substantially introduced by the 1947 Town and Country Planning Act, which vested development rights in the state, and by such legislation as the New Towns Act of 1946. Three subsequent unsuccessful attempts were made to effectively nationalise development land. Under Conservative administrations since 1979 the political rhetoric has indicated a decline in state intervention, but reality does not always bear this out.

The legislation changes from time to time, but overall the postwar years have been characterised by a combination of fiscal and regulatory approaches. During the 1980s there was a move towards more active, albeit highly selective, intervention designed to stimulate and support the market in problem areas.

Land policy is overseen by the Secretary of State for the Environment who issues statements and guidance through ministerial circulars and planning policy guidance notes. Currently, the main urban goals are inner area renewal and the containment of urban sprawl. Land initiatives are seen as a means to an end, not an end in themselves. At a local level much, but not all, land policy operates through county structure plans which take a medium term (10–15 years) view of broad strategic issues. Britain's counties are amongst the largest, but weakest, local government units in Europe and normally there is also a second tier of district councils which deal with many of the day-to-day issues, including, in some cases, the preparation of local plans. In London and the other former metropolitan counties, structure and local plans have been replaced by new unitary plans since 1989, and a debate on the reorganisation of other local government structures is currently gaining ground. By the mid 1980s, local authorities were reacting differently to the national framework according to their local needs and circumstances. In addition to the overall framework, many had a range of additional grants, programmes and agencies which they could bid for, or which were in some cases imposed upon them by central government. Even so, many of them found that their access to both finance and legal powers to control their local development was being reduced (RTPI 1985). Local planning and land policy was seen as being given a firm push in the direction of less public sector control and more encouragement for private sector activity.

West Germany

Leaving aside the uncertain consequences of unification, West Germany shows some similarities with Britain, but also many differences. It is a mixed economy with well developed and fairly harmonious public and private sectors, and it has several land policies which may be likened to more efficient versions of familiar British ones. German policies have been thinly documented in English, but Hallett (1977; 1988) is a notable exception. It is the federal government which establishes broad spatial patterns of economic and demographic development, and the Lander authorities then prepare regional policy and land development programmes. Detailed land policy is decided mainly at commune or city level, being controlled by a master plan of development which describes basic land use, together with a building plan for a few developing areas of the community.

The communal or municipal government plays a very active role, especially in ensuring the provision of land for future building through a long established programme of land banking. Most urban development land however remains in private hands and municipal governments have a declining role in this respect. Other responsibilities of local government include replotting, that is, the exchange of plots to ensure larger scale and better co-ordinated development, the administration of laws to tax profits on land sales, and the reactivation of disused or derelict land. The latter is also achieved through funds applied from the Lander through special agencies, for example the North Rhine–Westphalia Real Estate Fund and the Ruhr Real Estate Fund. Hallett (1988) has argued that West Germany has carried out some of the most successful urban renewal schemes and he cites the active participation by local authorities in the land market, and the use of quasi-public agencies responsible to the local authority as key features. Another major success of German planning and development policy is the housing structure. Owner occupation rates are lower than in Britain, but there is an active private rental sector and a generally popular and well regarded social housing sector. A particular feature in its success is that patterns of finance and land development have encouraged a small-scale, fine grained form of development (Hallett 1988) which avoids the obvious and stigmatised division between large public and private housing estates so often found elsewhere.

Land and property are subject to a number of different taxes in Germany, including a property tax (*grundsteuer*) similar to the domestic rating system abandoned in England in 1989, a wealth tax and income tax on profits from land sales.

France

Two important characteristics have helped to shape the French approach to land policy. The first is a large measure of centralised state control – perhaps a necessity in a country with over 36,000 local government units, 85 per cent of which have fewer than 1,500 people. The second is the relative recency of large scale urban growth, especially when compared with the UK, Germany or the USA.

Since 1946 a series of five year national plans has been produced in which land policy and town planning have been integrated with regional and national plans. One of the principal aims of land policy has been to ensure an adequate supply of housing land for the rapidly urbanising population. To this end, an approximate equivalent of the British Structure Plan exists in the form of the Schema Directeur, a large scale, 30-year urban development plan for major cities. At a more detailed level the *code de l'urbanisme* is an elaborate document governing planning and land development matters through a complex network of local government bodies and quasi-public agencies. Nearly all developments of more than a handful of houses involves public bodies and an array of instruments which remove most disputes over betterment.

The principal tool, and most distinctive feature of French land policy is, and has been, a number of clearly designated urban development or management zones, in which the state has various pre-emptive rights to purchase land. The most important of these in the postwar years have been as follows.

ZUP (Zone à urbaniser en priorité; 1958–76). These were zones designed to enable the streamlining of land acquisition by the local authority (commune), normally by means of funding from the state, in order to promote large scale housing development. Over 150 ZUPs were created, but owing to changing circumstances, and some public disillusionment, they were extinguished in 1976. The result of many of the ZUPs, especially with their socially segregated *grandes ensembles* of housing blocks was not dissimilar to the peripheral local authority housing estates of many British cities.

ZAD (Zone d'amenagement différé; 1962–present). The main purpose of these was to designate zones for future development, within which local authorities would minimise speculation by having pre-emptive rights to purchase land at the prices existing one year before designation.

ZAC (Zone d'amenagement concerté; 1967–present). In effect the ZAC was a replacement for the ZUP, in which largely residential development could be shared between the local authority and private developers. A particularly effective instrument in this is the joint private/public companies, the *sociétés d'économie mixte*, which enjoy some public powers, such as compulsory land purchase, but which are also firmly based in the funding and risk taking ethos of private business. Within the designated area, a comprehensive framework is established to acquire land, organise finance, provide infrastructure and undertake the actual development.

ZIF (Zone d'intervention foncière; 1975–present). These zones were a product of the Urban Land Law of 1975 which attempted to control rising land prices and speculation, and to encourage local authorities to provide more social housing, public open space and conservation measures in development and redevelopment schemes. As in the ZADs, the local authority has the power to purchase land at the previous year's prices. By 1983, 1,676 ZIFs, covering more than half a million hectares had been designated, but after 1986 they were no longer compulsory.

As in the UK and elsewhere, a pro-development stance became prominent in France during the 1980s. The return of a more right wing government in 1986 led to some relaxation of land policy, but there remains a strongly centralised control. Punter (1989) has provided a detailed account of French planning and land policy.

Italy

Although Italy has faced some of the most profound economic and social restructuring, and some of the most challenging land policy issues in postwar Europe, the national situation has been poorly documented. Land policy and planning, such as it was, was dictated by the State until 1970, but since then the regions and communes, which have always had considerable administrative power, have been gaining strength. In Italy, even defining the locus of the state introduces complications, for the administration and legislature is centred in Rome, but it is Milan and other nearby cities which have provided the real engine of economic and urban growth. It is thus in the north that much of the real conflict over land has arisen. This conflict, not surprisingly, has an intensely political flavour. On the one hand is the need of a developing urban industrial society to have a clear framework of land development and regulation policies, a need backed by the labour movement, and on the other hand lie the interests of the ruling party whose power is largely vested in real estate, development and speculation (Calavita 1984).

Much of the very rapid urban growth in Italy in the period 1950–70 took place without land use controls. The society that had built some of the most impressive towns in history was now making a poor job of urban growth. Development was uncontrolled, building standards were poor, infrastructure was sometimes absent and land and housing prices were rising rapidly. Speculation was widespread and many large corporations, whatever their primary activity, became widely involved in land and property dealing.

Urban land policy and planning had existed since the 'Urbanistic Law' of 1942 which laid down the need for master plans and land use plans and which gave local authorities the power to secure land for development at existing use prices, but the legislation was not universally applied. In 1962 local authorities were permitted, by Housing Law 167, to acquire land for low cost housing at, or below, market price and a number of them began to establish land banks. But even this skeletal form of policy was only partially implemented before the Bridging Act (765) of 1965 began to extend and enforce it.

The biggest change came in 1977 with Law No. 10, a new planning and land policy law. This began to tighten up the land development process by stipulating that any development needed a 'concession' from the local mayor. Such concessions would normally cost between 5 and 20 per cent of the project's construction costs, and amounted in effect to a development land tax. At the same time, development was to be concentrated within designated implementation programme areas, and within these areas land owners would have to develop their land or see it appropriated by the local authority. This law is equivocal and messy (Calavita 1984), but it has profoundly modified not just the development process, but also the whole structure of property rights in Italy.

The United States of America

Land policy in the USA has generally been guided by the importance of private property rights and the operation of the free market in motivating individuals and groups, but also by the recognition that, in specific situations other individuals may be adversely affected. These notions were enshrined in the balancing concept of the Supreme Court in the 1920s. This involved balancing the rights of individuals to be protected from adverse effects of development against the rights of property owners not to have their rights unduly restricted. Generally, however, subject to certain safeguards, US land policy and planning adopts a very positive attitude towards development.

Certain waves of activity may be identified; in particular, the vast land settlement programmes of the last half of the nineteenth century, the rise of detailed zoning ordinances in the 1920s and the wave of urban renewal, environmental awareness and energy related issues from the mid 1960s onwards. Today, urban land use and development policy is an active issue in many cities, and public support seems high (Dunlap 1987), but the country lacks a coherent and explicit land use policy framework (Dowall 1989). Policy has evolved in a relatively unguided way, through a myriad of federal programmes and uncoordinated local and state approaches. It has changed greatly since 1970, when it consisted almost wholly of local zoning, into a system which today is increasingly centralised at state, regional or federal levels (Popper 1988).

The federal role was small until recently, although it is true that highway development and mortgage guarantees were important, if indirect, aspects of land policy in the 1960s. Under the Carter administration an 'Urban Policy Agenda' was launched in the late 1970s, but it was repealed in the Reagan era. In the 1980s, federal interest in land development issues ran parallel to some aspects of government attention in the UK, being concerned with the deregulation of private enterprise, economic development, job creation and the revitalisation of run-down urban communities.

At state level intervention was also fairly minor until the 1970s, being mainly concerned with road programmes and economic development, but rapid urban growth and the environmental movement awoke interest. There arose a 'Quiet Revolution in Land Use Control' (Bosselman and Callies 1972), and by 1975, twenty new environmentally based state land use laws were passed. Florida, Rhode Island, Vermont, New Jersey, Maine, Delaware, and Georgia all have new legislation to require participation in state planning processes (Meck 1990). In recent years, however, many states have passed few new land laws, and those which exist have been weakened. Most states have chosen not to get involved directly in local land market regulation, but a few, including California and Massachusetts, do have active low-cost housing policies.

It is local government which is the main level of land policy and regulation, particularly with respect to zoning controls which remain the basis of US land policy, subdivision rules, the provision of infrastructure and the important business of bargaining with developers in order to maximise community benefits. A particularly lucid account of these processes and development control planning in the USA has recently been provided by Wakeford (1990). It is local government too which decides what land uses to permit, what taxes to levy and what services to provide. The fact that so much activity is concentrated at this level, coupled with the American retention of old and fragmented local government units, means that there is great diversity of policy, but the local emphasis means that it can also be more innovative than in other countries. Cities and counties in most states now prepare and update community master plans outlining population, economic and land use patterns, but there is a dichotomy between programmes designed to limit growth and those designed to promote it. A particularly comprehensive form of planning is seen in Oregon where local governments must prepare comprehensive land use plans in compliance with nineteen statewide goals and guidelines, some of which are in conflict with each other (Knapp 1987). The main elements of this are home rule, urban growth management, economic development and housing. Elsewhere, large cities with declining local economies have been active in public/private partnerships and other measures to revitalise downtown areas. For example, San Diego, Baltimore, Oakland and St Louis have all provided land and low cost financing to attract developers. In some depressed urban areas, land banking, tax holidays, land write downs and urban development action grants are used.

Elsewhere, controlling, not promoting growth is the problem. For example, in Boulder, Colorado, there is a limit on annual building permits and in a number of Californian cities, including San Francisco and Los Angeles, there are other forms of growth limiting regulations. The requirement for developers to demonstrate the adequacy of public infrastructure, or to fund new services, is becoming increasingly common. In areas of high growth, concern focuses mainly upon the problems of overburdened roads,

environmental impacts, the transformation of formerly quiet neighbourhoods into highly urbanised areas, the lack of affordable housing and the
question of who pays for infrastructure. Occasionally, adjacent local authorities will act together to create a 'suburban-squeeze' (Dowall 1989), as in the
San Francisco Bay Area, Washington DC, Montgomery County, Maryland,
and Fairfax County, Virginia. In one of the most substantive reviews of the
subject (de Neufville 1981) American land policy was seen in terms of four
components: economic, social, environmental and capitalist.

1   Land use policy as economic policy involves promoting growth, main-
    taining economic stability, using resources efficiently and redistributing
    wealth. Equally it can be interpreted as an unintended effect of other
    policy fields; for example, the location of defence spending, energy poli-
    cies and transport decisions all have land use impacts.
2   Land use issues are inevitably tied up with social policy, albeit normally
    at an implicit, rather than explicit level. For example, homogenous low-
    density suburbs have become a major outcome of zoning ordinances,
    but they have not been part of an explicit national land use policy. In
    particular cases, an emphasis upon economic objectives, for example,
    downtown redevelopment, has tended to ignore social problems in the
    wider city area (Keating and Krumholz 1991).
3   Environmental policies and concerns have been a strong influence in the
    past twenty years, encompassing a variety of ecological notions, carrying
    capacities, conservation themes and aesthetic considerations. One major
    criticism levelled at American planning by Delafons (1991), in an other-
    wise favourable review, is that it has tended to regard land as an inex-
    haustible resource.
4   In the capitalist context land use policy has been seen by Markusen
    (1981), Boyer (1981) and others, not so much as a set of good ideas to be
    applied for a tolerable social order, but more as a consequence of the
    contradictions within capitalism. Thus state intervention is prompted by
    the need to achieve legitimation, or to minimise class conflict and social
    injustice. In this way, land use controls will be weak during a phase of
    growth of capital but will increase during a period of economic decline.

Any judgement on the evolution of American land policies in the recent past
is bound to reveal contradictions. The pattern of physical development has
clearly become very uneven, and at a local level there has been a great
diversity of both regulatory and promotion policies according to circumstances. Numerous programmes have emerged at state and local level, but
they do not really add up to a consistent or coherent package, and there are
no nationally agreed land use goals. That land use regulation has changed
since 1970 is fairly clear. In 1970 it consisted almost entirely of local zoning,
but since that date there has been increased regional, state and even federal
intervention. Popper (1988) suggested that two alternative interpretations are

common. The liberal view is that planning necessarily became more extensive and more centralised after 1970, with state and federal bodies trying to
overcome environmental shortcomings of the skeletal zoning system. The
conservative anti-regulatory onslaught and the fiscal austerity of the 1980s
meant that they never had a chance to succeed, and, in particular, the
prospect of federal land use regulation all but disappeared. In this, the failure
of Congress to pass a National Land Use Policy Act, which would have given
the whole country laws regulating large development, similar to those in
Florida, Oregon and Vermont, was particularly significant. The alternative,
conservative, view is that the new initiatives amounted to an over-reaching
burden of bureaucratic regulation. Conservatives cited the growth of legislation such as the Clean Air Acts, Clean Water Acts and Surface Mine
Control Acts and the cumulative effect of state and federal agencies which
meant that 'private land is now one of the most centrally regulated sectors of
the American economy' (Popper 1988: 295). Popper's own explanation was
neither the Conservative nor the Liberal one. He argued that land use regulation did not collapse, but that it has continued to expand, but quietly and
in a different form. He suggested that there is more centralised control than
ever, but it is specialised and geared to particular purposes, and it does not
represent a comprehensive or coherent policy.

As in Britain, there has been a marked trend towards short term operational effectiveness – what Meck (1990) described as a new pragmatism.
Also, again as in Britain, there has been an important trend towards the privatisation of certain government functions, but with many cities moving in
the opposite direction by de-privatising real estate development. Many
public agencies of various types are now involved in deal-making partnerships with private developers, especially in downtown or waterfront revitalisation schemes (Frieden 1990).

CONCLUSION

All governments accept the need for some form of intervention in the urban
land development process, although this varies from the minimal and piecemeal measures found in the USA, to relatively high and co-ordinated levels
in The Netherlands and Sweden. In many countries land policy is fragmented, with land ownership, land values and land use being treated separately. In addition, it can be suggested that the past two decades have been
particularly challenging ones for land policy makers. Many urban economies, and their land use patterns have seen a more profound restructuring
than at any other time in the past century. In a number of cities this has been
not simply another economic slump of the kind which they have experienced
before, but a complete decline of major areas of industrial activity and a
transition into a new and uncertain economic order. A major manifestation
of this, in some countries, has been the creation of widespread areas of vacant

and derelict land, a theme which will be taken up in the next chapter. The same time period has also witnessed a movement of government policies away from state intervention, notably in the UK and the USA. At times this has produced contradictions whereby governments have been professing a policy of non-intervention whilst actually manipulating some aspects of the land market in more detail than ever before.

Land policy, particularly where it involves significant state intervention, is frequently controversial, no matter how laudable its aims may be. Some of the difficulties have emerged in the discussion above, but it will be useful to draw together the main issues by way of summary.

1) *Objectives*. Land policy rarely has a single objective. Multiple aims are common, as in the case of land taxes which may be intended to raise revenue, regulate the market and send out social/ideological signals to the community. But multiple aims may conflict or even be incompatible. Social equity and economic efficiency are rarely easy to reconcile.

2) *Negative side effects*. It is relatively easy to see that some kinds of public intervention, designed to regulate development, may in fact result in a slowing down or reduction of development. For example, development land tax may reduce the incentive to develop land, or at least affect the timing. Taxes on development land may lower the price or profit received by the seller, but they do not normally lower the price or increase the availability for the buyer.

3) *Practicalities*. It is often difficult to turn sound conceptual reasons for public intervention into robust workable policies with no unforeseen side effects. This is particularly true in the case of complex development schemes covering large areas where it is virtually impossible accurately to assess the full picture of losers and gainers and hence who should be taxed and who protected or compensated.

4) *Loopholes*. Most policies, inevitably, have loopholes, or can be used for purposes for which they were not originally intended. For example, it has been argued that developers were able to take advantage of the special incentives and generally pro-development ethos of the 1980s in order to increase their profits or to by-pass planning regulations. Similarly, private individuals may profit financially from selling houses bought with government subsidised mortgages.

5) *Political structures*. Leaving aside for one minute the ideologies of governments, there are certain political structures which greatly influence land policy. In Britain a sharp rift has been exposed between a central government dedicated to the operation of market forces, and left-wing local authorities within whose areas some of the most problematic land issues are to be found. In American cities too, notably New York, there has been much tension between city hall, developers and local community groups. The role and influence of civil servants is also important. In France, for example, they are both more centralised and more interventionist than in Britain, although

the situation has changed as promotional policies in Britain have become more active and more centralised. In comparing European political structures it can be suggested that Britain's adversarial two party system has resulted in wide swings of land policy and it has inhibited the development of a coherent and consistent programme.

7

VACANT AND DERELICT LAND

Since the early 1970s, the problems of vacant and derelict urban land have become increasingly obvious in a number of cities. It is not a universal problem, some cities and some governments have either avoided its worst incidence or have dealt with it successfully, but in other urban areas, it has been, and remains, a pressing land use issue. The areas which have been worst affected include large, old cities, especially where heavy industries and dock areas have become run down and where the local economies have been insufficiently dynamic or flexible to absorb the land into new uses. Nationally, it is the cities of Britain, and, to a lesser extent, of North America which have been worst affected, but those in France, Germany and even Australia have not been immune. In Britain, dereliction and vacancy have been an intrinsic part of the inner city problem and the condition has been seen as both a symptom and a cause of wider urban problems. It has political and social dimensions, but above all, it came to symbolise the decline of old urban economies in the two decades after 1970.

THE EXTENT AND NATURE OF URBAN WASTELAND

The extent of vacant, derelict and otherwise unused land in the major cities increased rapidly at the time of the economic recession prompted by the oil crisis in 1973. Soon large tracts of the inner city consisted of little more than derelict land and buildings, representing substantial holes in the city (Dawson 1979). By the mid 1980s, it was suggested that 'wasteland is probably the most characteristic denominator of these areas' (Moor 1985: 56). Sample local authority surveys suggested that, in the mid 1970s, on average, 5 per cent of land in metropolitan areas was vacant (Burrows 1978). For a few boroughs in the East End of London, and for inner Glasgow and Liverpool, the vacancy rate was over 10 per cent.

Comparative figures showing the extent of wasteland are difficult to obtain even within Britain, let alone internationally. A main reason for this is the variation in definitions and methods of collecting data. In England, derelict land (sensu stricto) has long been defined as 'land so damaged by industrial or other development that it is incapable of beneficial use without treatment' (DoE 1991: 2), but this is a restrictive definition which excludes much land that is effectively unused. A recent suggestion that the definition be changed to 'land significantly damaged by industrial or other use' (DoE 1989: 3) would widen the scope somewhat. Vacant land, in the British planning context, is defined more widely. For the purpose of compiling the Land Registers of publicly held vacant land which have been required since 1980, the definition is taken as land which 'in the opinion of the Secretary of State ... is not being used, or is not being sufficiently used for the purposes of the performance of the body's functions of carrying on their undertaking'.

Using these definitions, official figures gave totals of 45,683 ha of derelict land in the 1982 Survey of Derelict Land in England, and 43,550 ha of vacant land recorded on the Land Register in 1983. Both figures were clearly under-statements of the overall problem. In the case of derelict land, under-estimation occurs for a number of reasons. Amongst these are: the exclusion of unused land which lies within the boundary of an otherwise active under-taking, the imposition of a minimum site size criterion, the exclusion of sites which are unused but have planning permission for future development and the fact that local authorities may classify as derelict only those sites on which there is a good chance of them receiving a Derelict Land Grant for restor-ation work. Independent estimates put the figures considerably higher, with some agreement around a figure of 210,000 ha for the combined total of derelict and vacant land (Chisholm and Kivell 1987).

The most recent official survey of derelict land in England is that compiled from local authority figures in 1988 (DoE 1991). This revealed a total of 40,500 ha, equally divided between urban and rural areas. When compared with previous surveys, this set of figures suggests that the total amount of dereliction has, for the first time, started to decline. The drop between 1982 and 1988 was 11 per cent. The downward trend is confirmed by figures from the declines in vacant land recorded by annual land use change statistics (DoE 1988). This source showed that 3,010 ha of previously vacant land had been brought back into use in the previous year, compared with 1,450 ha of newly created vacancy, a net fall of 1,560 ha.

The incidence of different types of dereliction is shown in Table 7.1. To some extent the divisions are arbitrary, but the table indicates that almost half (45 per cent) of the total in 1988 took the form of spoil heaps and exca-vations, and was attributable to mineral extraction. Approximately one-third (31 per cent) was classified as general industrial or 'other forms', covering a broad spectrum of manufacturing industry, public utilities and dockland closures. Ownership is also an important part of the derelict land issue, and the same survey revealed that, where ownership could be established, the overall split between the public and private sectors was equal. However, in urban areas, the public sector, mainly local authorities, owned 60 per cent of derelict land.

151

Table 7.1 The amount of derelict land and the proportion justifying reclamation, by type of dereliction, April 1988

Type of dereliction	Area (ha)	% justifying reclamation
Spoil heaps	11,900	63
Excavations and pits	6,000	73
Military etc. dereliction	2,600	80
Derelict railway land	6,400	79
Mining subsidence etc.	1,000	90
General industrial dereliction	8,500	94
Other forms of dereliction	4,100	92
Total	40,500	78

Source: Survey of Derelict Land in England 1988, Department of the Environment 1991

The location of dereliction in 1988 showed a marked regional bias, with over half (52 per cent) ocurring in the three most northerly regions, reflecting the industrial and mining origins of much of the problem. Large cities accounted for a disproportionate share (Table 7.2) with the seven major conurbations containing over one-third of the total, and 46 per cent of the industrial and 'other' categories. Outside of these conurbations, a few other urban districts, notably Stockton and Langbaurgh (Cleveland), Hull and Stoke on Trent also showed large concentrations.

Two important trends can be identified which have particular relevance for urban areas. One concerns the type of dereliction. For most types of dereliction (mineral related, military and railway land), as Figure 7.1 shows, the total figures declined between 1974 and 1988, largely due to active programmes of reclamation. For other forms of dereliction (mainly indus-

Table 7.2 Derelict land in major metropolitan areas, 1988

Metropolitan area	Derelict land (ha)	Industrial and other (ha)
Greater Manchester	2,872	1,203
West Yorkshire	2,843	823
West Midlands	2,281	1,052
South Yorkshire	2,183	850
Merseyside	1,473	928
Greater London	1,386	556
Tyne and Wear	1,068	335
Metropolitan areas total	14,106	5,747
England total	40,495	12,613
% in Metropolitan areas	35	46

Source: Survey of Derelict Land in England 1988, Department of the Environment 1991

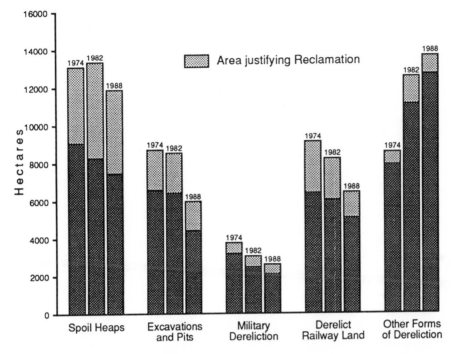

Area justifying Reclamation

Source: *Survey of Derelict Land in England 1988*. Department of the Environment 1991
Figure 7.1 Changes in the amount of derelict land and the proportion justifying
reclamation, 1974–88, by type of dereliction

trial), however, the totals increased. Despite 5,482 ha of reclamation in this
category between 1982 and 1988, the total stock actually grew by 1,109 ha.
Thus in the mid 1980s this kind of derelict land was being created at a gross
rate of approximately 1,000 ha per year. The second (related) trend is that the
proportion of the national total of dereliction found in the seven major
conurbations increased from 27 per cent in 1974 to 35 per cent in 1988. It can
thus be argued that, despite an active reclamation programme and a number
of local successes, there was a stubborn body of industrial dereliction in the
major cities which increased in both absolute and relative terms between
1982 and 1988. Only towards the end of the period were there some signs
that the net total was beginning to decline.

A rather different set of statistics is that relating to vacant land. As previously mentioned this is defined differently from derelict land, and although
there is some overlap, in 1988, only 14 per cent of derelict land was recorded
on the Register of Vacant Land (DoE 1991). In February 1987 a total of
40,235 ha of vacant land was recorded on this Land Register, equivalent to
four times the area of the city of Manchester . Again, the major conurbations
were the main focus (Table 7.3) collectively accounting for 28 per cent of the

Table 7.3 Vacant land on the public land register in major urban areas, 1987 (hectares)

Metropolitan counties	Area vacant 1987	Disposed of 1984–7	Brought into use 1984–7
Greater London	1,896	947	394
Greater Manchester	2,173	461	200
Merseyside	1,399	64	33
South Yorkshire	1,543	63	13
Tyne and Wear	1,657	263	130
West Midlands	960	328	91
West Yorkshire	2,002	113	20
Total	11,630	2,239	881
Urban districts over 200,000 population			
Bristol	353	60	13
Derby	205	43	4
Hull	524	41	23
Leicester	330	18	25
Nottingham	164	22	24
Plymouth	222	36	15
Southampton	38	6	9
Stoke on Trent	219	17	8
Total	2,055	243	121

Source: Department of the Environment Land Register, 19 February 1987

total. Recent changes in the organisation of the Land Registers mean that local authorities no longer have information relating to vacant land held by other public bodies. However, it is clear that the local authorities themselves hold by far the greatest part of publicly owned vacant land – Table 7.4. Individual totals can be substantial, for example in the Greater Manchester area in 1990, Manchester City Council owned 166 vacant ha, Tameside Borough 140 ha and Bolton 171 ha. It should also be noted that the 'public' status of a number of the bodies on this table changed as a result of privatisation programmes in the late 1980s.

Clearly derelict and otherwise vacant land must be considered a transitional phase, or a flow, within the urban development process (Bruton and Gore 1981; DoE 1989) whereby the total at any given time may be diminished by reclamation, or added to by new dereliction – Figure 7.2. Nicholson (1984: 20) saw it as a 'transient feature of the urban environment, brought about by economic and social changes stimulating adjustments in land use patterns'. Even so, there is concern over the long duration of vacancy in many cases. Chisholm and Kivell (1987) showed that a range of 10–15 years is relatively common and a recent survey of 375 sites (Civic Trust 1988) revealed that 78 per cent of them had been vacant for more than 5 years, and one third for between 10 and 25 years. In east Manchester, Baum (1985)

Table 7.4 Vacant land ownership, 1987

Owning body	Area (ha)	% of Total
District authorities	18,023	44.7
County authorities	5,981	14.9
British Rail	4,145	10.3
New town	3,249	8.1
DHSS	1,976	4.9
Electricity boards	1,758	4.4
Port authorities	1,283	3.2
Water authorities	906	2.3
National Coal Board	855	2.1
British Steel Corporation	486	1.2
Urban development corporations	452	1.1
Defence departments	409	1.0
Department of the Environment	331	0.8
Other	381	1.0
Total	40,235	100.0

Source: Department of the Environment Land Register, 19 February 1987

measured the mean period of vacancy at $5\frac{1}{2}$ years.

A rather more detailed view of one important facet of land and property vacancy can be obtained by examining the few, fragmented, studies of disused industrial floorspace. For England and Wales as a whole, Falk (1985) estimated that the decline of manufacturing had created 16.25 million square metres of empty industrial buildings in the mid 1980s. In the Black Country, Watson (1987) identified 1.6 million square metres of vacant industrial floorspace, of which he claimed that 14 per cent was obsolete and 40 per cent had little long term potential. A detailed study of Stoke-on-Trent (Ball 1989) identified 146,000 square metres of vacant industrial premises in 1987 although the figure had halved in the previous two years. Old buildings (pre-1918) and those constructed for specialised industrial purposes were particularly vulnerable. At the depth of the recession in 1982, a survey by property consultants (Thorpe and Partners 1982) identified a total of 3.1 million square metres of vacant industrial floorspace in the 32 London boroughs, much of it in new premises. Five boroughs had between one-third and a half of their freehold industrial property vacant. Unused property, as well as unused land had thus become an endemic feature of Britain's manufacturing centres, and many of the reasons for this will be discussed later in this chapter.

Outside of Britain, derelict and vacant urban land does exist, but it occurs in more localised forms and rarely becomes the extensive and enduring problem that it is in many British cities. National statistics are virtually non-existent, but it is clear that, of our immediate neighbours, both France and Germany have comparable individual examples of dereliction but a far less

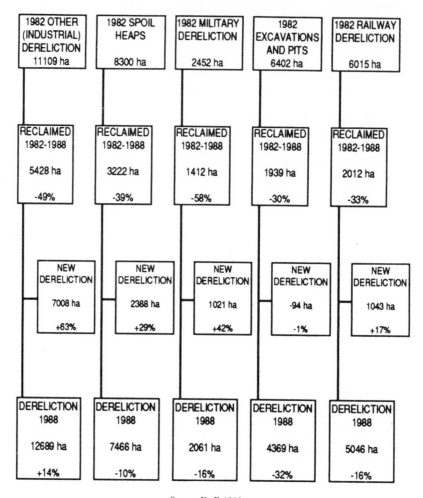

Source: DoE 1989

Figure 7.2 Flows into and out of dereliction, by type

severe problem overall. In France the problem of *friches industrielles* was investigated by the Lacaze Report of 1985 (Clout 1988). This found a total of 20,000 ha of dereliction, although, as in Britain the official figure is thought to be an underestimate. The industrial recession of the late 1970s and early 1980s was the worst period for the growth of dereliction and its rate of creation is now thought to be declining. Three kinds of derelict/vacant land were identified:

1   Land abandoned by the user, but quickly taken up by another with no

state involvement. This is common in areas of high demand like the Ille de France.

2 Land where demand for reuse exists, but at such a low level that state help is needed.

3 Land where demand is so low that public open space or agriculture is the only possible reuse. This is common in the Pas de Calais.

The problem is mainly concentrated in three regions (Couch 1989), and especially in the Pas de Calais area where 1 per cent of the land is derelict (Ernecq 1988) and half the national total is to be found. Using the rather different definition of vacant industrial land as land empty for over one year, Malezieux (1987) recorded 9,400 ha on 1,200 sites in the Nord/Pas de Calais area, 1,647 ha in Lorraine, 477 ha in Rhone/Alps and 544 ha in the Ile de France region. Dereliction in the Ile de France was concentrated along the axis of the Seine and among industrial districts in the north and north east of Paris. Unlike the single industry dereliction of Pas de Calais and Lorraine, that in Paris results from a complex restructuring of the urban economy involving old canals and railways as well as a variety of industries. A sample survey of Paris (Chaix 1989) found that industrial dereliction was absent from over half the communes and overall accounted for less than 5 per cent of industrial land. Vacant land gets reused relatively quickly in France. For example, between 1984 and 1987, one-third of the 1984 total of derelict land in Paris was redeveloped, two-thirds of it for economic activity (Chaix 1989).

Elsewhere in Northern Europe, the processes of industrial change have also contributed to the creation of dereliction and temporarily abandoned land. Particularly hard hit has been the Ruhr district of Germany where a generation of coal and steel closures had created over 2,000 ha of dereliction by the late 1970s (Couch and Herson 1986). In this area the problem has been worsened by a fragmentation of ownership and the inability, or disinclination, of private owners to restore derelict sites. Similar problems are to be found across a wide swathe of declining coal and steel disricts in Germany, The Netherlands and Belgium (Aitken 1988). The other main circumstance in which derelict land occurs is in association with dockland changes, as for example, in Duisburg, Hamburg and Rotterdam.

Apart from industrial dereliction, the urban development process itself, and associated speculation, results in vacancy. Sinn (1986) estimated that approximately 10 per cent of existing lots in German cities were either vacant, or were being used for inferior uses, but he saw this as a requirement of an efficient land market, not as a sign of market failure.

In the United States the pattern of urban dereliction and awareness of it is again different. This results from differences in the timing and form of both industrial and urban development, from a dissimilar system of planning and public intervention and from a rather different set of market conditions. Broadly, the issue of derelict/vacant land is not a pressing urban problem in the USA, but there are some significant local concentrations. Single cause dereliction resulting from coal mining is common in parts of Pennsylvania and West Virginia, and much land has been damaged by the oil industry in Louisiana, but these are not especially urban problems. In cities, dereliction and vacancy occurs mostly as a result of either urban restructuring, including general inner city changes and waterfront development, or as a result of processes of urban growth which often leave behind vacant and undeveloped plots. In some cities the decline of the inner districts, especially in the period between 1950 and 1980, was associated with widespread land and property abandonment. In Philadelphia, for example, in the mid 1970s, there were estimated to be 26,000 vacant and abandoned houses and 12,000 vacant lots (Bacon 1976).

Many cities in the United States have been taking stock of land left vacant by the development process. In Dallas, 526 ha of vacant land zoned for residential use was recorded in the inner city in 1970 (Dallas undated) and in Portland, Oregon, vacant land was one of the largest categories in the city, accounting for 17 per cent of the total (Portland 1978). In this case, constraints such as steep slopes, potential flooding and lack of infrastructure explained much. Across the nation, vacant land parcels were estimated to comprise one fifth of land in big cities (Northam 1971), and the total value of vacant, but buildable, land in a sample of eighty-six large cities was put at $6 billion. Recently, sophisticated computerised inventories have been used to keep track of vacant land. In Cincinnati, for example, where economic development is being heavily promoted, a scheme known as Site Finder is used to log details of vacant industrial/commercial land. In 1985, 600 vacant parcels, totalling 567 ha were listed (Carlsson and Duffy 1985).

In summary, it appears that derelict and vacant land occurs widely in the cities of the western world and in some of them it is a particularly concentrated and enduring problem. For the most part, other countries appear to be less badly affected than Britain and three explanations for this can initially be offered:

1 Britain was the first nation to experience large scale industrialisation and urbanisation and this has left it with a larger legacy of older plants and older locations than most countries.
2 The British economy was one of the hardest hit by the international economic recession of the 1970s and 1980s. Much of its manufacturing industry was vulnerable and widespread closures ensued.
3 A number of factors unique to the UK have played an important role. Many of these have been discussed elsewhere by Chisholm and Kivell (1987) but a few examples will be helpful here. They include the planning system which effectively grants land use rights in perpetuity, and thus encourages land owners to gear their market price expectations to some historic pattern of activity rather than to current realities; a tax

regime which levies no rates or taxes on vacant land; housing clearance policies operated by a number of local authorities in the 1960s and 1970s in which the rate of demolition and clearance far exceeded their capacity for rebuilding.

Superficially, the problems presented by derelict/vacant land are easy to identify, although there are also a number of other less obvious ramifications. To the general public, the issues of safety and aesthetics are perhaps the most obvious. Land which is visually ugly or damaged alienates the local community and it should not be forgotten that the policies of derelict land reclamation in Britain were pursued far more vigorously following the large loss of life caused by the collapse in 1966 of a colliery tip at Aberfan in South Wales. Aesthetics and safety sometimes go hand in hand, as with unstable tips or flooded excavations, but frequently the greatest threats to safety are hidden, in the form of potential subsidence, concealed shafts or toxic substances.

To planners and politicians, safety and aesthetics are important, but economic aspects of dereliction also claim their attention. Vacant land represents a waste of a resource, both in the sense that it is used for no productive purpose and because, under most systems of taxation it will produce no revenue for the community. Derelict/vacant land may also have an economic impact beyond the limits of its own boundaries in blighting surrounding sites and deterring development. In this sense the scale of the problem is important. One derelict site in the middle of an otherwise dynamic area may have little impact, but a locality in which dereliction is a characteristic feature of the landscape will find it difficult to start, let alone to sustain, the process of regeneration.

Attitudes also vary according to the scale and nature of the dereliction, but whereas in the UK it is normally seen in negative terms, as a problem to be tackled, in France (Thomas and Cretin 1987) and in the USA (Fox 1989), where it is less pervasive, it is more likely to be seen as a positive opportunity for reshaping cities and providing for community needs. In parts of America it has been seen as the new frontier for real estate developers and landscape architects.

CAUSES OF VACANT AND DERELICT LAND

Many explanations of vacant and derelict land are possible and for any given site a number of factors, both general and local, may be operating in concert. The general explanations help to explain the overall picture, but given that they include such broad factors as the state of the national economy, the functioning (or malfunctioning) of the urban land market, the processes of urban decentralisation and the operation of public policy in such fields as employment, housing and transport, then their particular roles in accounting for derelict and vacant land become both confused and contentious.

In a substantial review of the problem of vacant land, Cameron *et al.* (1988) traced three intellectual antecedents to the issue; 1) orthodox neoclassical economics, which sees the problem in terms of market failure; 2) the process view, which sees vacancy largely as a transitional phase and not therefore a major problem; and 3) the structural view, in which vacancy is identified as a deep seated problem of the postindustrial economy of Britain which can only be cured by broad structural changes. They went on to distinguish between factors causing the previous land use to cease, factors causing vacancy to continue and factors preventing temporary uses from arising. These distinctions provide a valuable framework, but for the present, briefer consideration, a simple division into general and specific causes will be adopted.

General causes

Structure and location of the urban economy

The past two decades have undoubtedly been a period of profound transformation for many of the urban industrial economies of Europe and North America. In confronting a postindustrial future, many cities have experienced far reaching economic, technological and social changes. These have, in turn, caused readjustments in land use patterns, many of which were originally formed a century ago, and the creation, perhaps only temporarily, of much vacant land.

At the heart of the transformation has been the economic recession which depressed demand and business confidence very widely for several years. But overlaid on this has been a more specific industrial restructuring and modernisation which was overdue in many older urban areas. Major cities and industrial regions have seen their markets disappear, or taken over by more efficient overseas manufacturers, especially in such fields as shipbuilding, textiles, engineering, metal manufacturing and vehicles. Some of these areas have been tipped suddenly into a postindustrial age in which large quantities of labour, social institutions and land have been made redundant.

Macro-economic changes and their impact upon land use patterns have been widely discussed (e.g. Massey and Meeghan 1982; Martin and Rowthorn 1986) and the notion of redundant spaces in cities has been explored by Anderson *et al.* (1983). The latter work set the context of industrial decline and social transformation very fully, but ironically said little about vacant land per se. It argued that uneven development is a specific characteristic of capitalism and that urban and regional 'industrial graveyards' develop. The concept of core–periphery is invoked to help explain the performance of different localities. Similar themes were raised by Rose (1986) who stressed that new roles need to be found for older manufacturing cities faced with the problems of adapting to social and economic change. In these adaptive processes, some cities are evidently coping better than others, for reasons which, as Rose suggested, are not entirely clear.

The link between industrial decline and vacant land is a fairly direct one, as anybody who travels across such cities as Manchester, Sheffield, Birmingham, Pittsburgh or Detroit can see. Closed factories lead to vacant/derelict land and buildings. The link is also deeper than this because the decline of manufacturing brings a decline in jobs, a loss of economic power and momentum and a number of other negative multipliers. The city then becomes less attractive as a residential and social environment and people move out. The processes of suburbanisation and urban decentralisation become powerful general causes of vacant land by lowering inner urban population levels and market demand.

But the argument is not simply about urban decline, it is also about restructuring and transformation. At the broadest level this includes the contentions of Lefebvre (1970), Harvey (1974) and others that a major part of the reorganisation of western urban economies has involved a shift from the primary circuit of capital, where industrial investment and economic growth predominate, to a secondary circuit where consumption and the power of finance capital dominate. Thus there are cases where manufacturing has not collapsed, but has restructured and relocated away from multiple inner city plants and locations to new, urban fringe locations. Similarly, changes in transport technology, especially the replacement of rail and water modes by road and air, have contributed to the locational obsolescence of old sites.

The role of industrial decline in creating vacant land is a particularly powerful one for it acts on both the supply and demand sides of the equation. Individual industrial closures create vacant sites, and the aggregate effects of decline both lower the local demand for land and contribute to a blighted and unattractive environment, into which new industries are unlikely to move. This is particularly true of many of the high-tech growth industries. The personnel director of a large American computer company put it bluntly by suggesting that sites throughout most of Britain's manufacturing heartland would be suicidal because of the difficulties of luring bright people to work in places with bad images (*Financial Times*, 12/11/86).

In the final quarter of this century a number of cities are having to adjust their space economies to the needs of the next century, whilst coping with a land use skeleton which was established in the last century.

Malfunctioning land markets

Markets malfunction for several reasons, and it sometimes happens that attempts to correct perceived difficulties result in further distortions. In any case, land as a commodity shows many peculiarities and the market is traditionally subject to cycles of boom and slump.

The role of market malfunctions in causing or maintaining vacant land has been widely debated. Supply side deficiencies, including artificially high land prices, sites being too small for development or unfit for use, high restoration costs, ownership and planning constraints and rent controls have been discussed by Coleman (1982) and, more extensively by Chisholm and Kivell (1987). Cameron *et al.* (1988) also referred to many of these, but they tended to emphasise demand failures. Clearly, in discussions of the role of the market there is as much scope for political as economic comment. Those who favour high levels of state intervention, collectivism and relatively traditional forms of land use planning see the vacant land problem as evidence of market failure. They cite high land acquisition and restoration costs, the speculative behaviour of land owners and developers, the fragmented nature of ownership and the absence of any obligation upon owners to reclaim derelict land as some of the key problems. On the other hand, those who subscribe to free market views see the main problem as the progressive handicapping of the market by public sector intervention. In particular, they point to the way in which the land market has been locked up by local authorities and nationalised industries who own large acreages and set unrealistically high base levels for land values, to the restrictive effects of land use zoning and to the existence of use rights which may remain fixed regardless of changing local circumstances. In short, a growing supply of land resulting from industrial and commercial closures has been accompanied by a falling demand due to low levels of business confidence either generally or in specific localities. Normally the market might be expected to adjust to these circumstances, notably through falling prices, but for the most part this has not happened.

In Manchester, Adams *et al.* (1985) suggested that valuation practices play a key role in preventing land prices from falling during a period of oversupply, thus causing a blockage in the development process. The widely used comparative valuation method is unable to cope with a declining market where few transactions take place. Prices are thus revised downwards only very slowly, if at all, the market fails to clear at anything like a full use of land equilibrium and vacancy ensues. In this context it is useful to reproduce (*pace* Partington 1986) some of the different concepts of land value and price which are well understood by valuers and surveyors, but by few outside of those professions.

- *Book value*: value of land/property as a capital asset as shown on the accounts, may reflect historic acquisition costs.
- *Existing use value*: open market value for existing use with vacant possession.
- *Open market value*: price likely to be realised if offered for sale in an open market after adequate advertising.

- *Market price*: price in a given market.
- *Hope value*: element of open market value over and above existing use value and reflecting the prospect of a more profitable future use.

Finally, it is apposite to return to the point that vacant land, per se, is not necessarily an indication of market failure, indeed some level of vacancy is necessary in a dynamic market (Sinn 1986). What indicates market failure in so many British cities is the volume and duration of vacancy and wasteland.

Ownership constraints

A key feature in explaining derelict and vacant land is the behaviour of the owners of the land. Land ownership in general was discussed in Chapter 5, and it is clear, especially from the work of Adams *et al.* (1988) that the role of the land owner, and the difference between passive and active ownership is important in bringing land into development. In the context of the inner city, they argued that development may be delayed (and vacancy therefore prolonged) by the unwillingness of a passive owner to sell, and this may be especially characteristic of the public sector. Additionally, fragmented or multiple ownership of a site may also complicate and prolong vacancy.

Within the overall pattern of ownership it is the public sector which frequently has been identified as a particularly guilty party. Certainly, many of the government measures designed to combat dereliction and vacancy in Britain, and the DoE studies upon which they were based, have targetted public landholders. Many studies (Dawson 1979; Chisholm and Kivell 1987; Loveless 1987; Cameron *et al.* 1988) point to the high level of public owner-ship of vacant land, and the 1988 Survey of Derelict Land in England showed that over 60 per cent of urban dereliction in known ownership was the responsibility of the public sector. A recent study by the *Sunday Times* news-paper (4/2/90) suggested that state owned bodies are often reluctant to dispose of surplus land and it listed the following totals of derelict/vacant land: British Rail 3,035 ha, British Coal 3,845 ha, British Steel 810 ha, British Waterways 810 ha, British Gas 610 ha, Ministry of Defence 1,015 ha. Clearly, various public bodies are substantial holders of wasteland, but it is also fair to point out that their totals are falling, whereas those of the private sector are rising (Adams *et al.* 1987; Civic Trust 1988).

Of course the mere fact of public ownership alone is insufficient to explain vacant land. The causal mechanisms which are normally cited are as follows:

1 Public sector bodies mismanaging or neglecting their land, through inefficiency, ignorance of what they hold or because land holding is incidental to their main activity.
2 The unwillingness of public bodies to sell land in a thin or falling market at less than its historic cost.
3 The holding of land for some future expansion.

4   The presence of a local authority as a 'buyer of last resort' may artificially keep prices up and distort the market.

Local authority policies

Closely related to the question of public ownership of land is the operation of
a number of local government policies which have contributed to land
vacancy almost uniquely in British cities. For the most part, there has been
nothing deliberate about this (although some local authorities have been
unenterprising towards development), but there have been a number of
housing, transport and land use policies which have undeniably created
vacant land.

Prominent amongst such policies were the grandiose slum clearance and
road improvement schemes which many authorities started in the 1950s, but
which in a number of cases were changed, delayed or simply not completed.
Much of the vacancy and dereliction in Liverpool, one of Britain's most
blighted cities, is due to these causes and the same is true for Glasgow and
Edinburgh (Dawson 1979). In Manchester, the very necessary slum clearance
schemes which demolished 83,000 dwellings between 1951 and 1981 were so
extensive and rapid that redevelopment could not keep pace and vast tracts
of dereliction resulted. In the six worst affected wards, one-third of vacant
land was due to housing clearance policies (Adams *et al.* 1987). Similar
examples can be cited from the field of transport planning. Again, in
Manchester, a 1962 SELNEC plan envisaged a series of ring roads and a
complex feeder network. Only 20 per cent of the plan was actually
completed, but clearance and planning blight created large areas of vacant
land. Changes of policy resulted in the abandonment of 178 km of new high-
ways in fifty-eight schemes around the city between 1971 and 1985 (Adams *et
al.* 1987). By the mid 1980s attitudes had changed and local authority
planning policies became a lesser factor in creating wasteland; indeed it was
more common for some authorities to be accused of almost indecent haste in
trying to promote any kind of development.

One further, and very vexed, aspect of local authority policies that
impinges upon vacant land is the question of local rates or taxes. Coleman, in
particular, has argued that high, and rapidly rising rates drive economic
activity out of the city. She demonstrated that industrial rates in Tower
Hamlets rose by 800 per cent in three years (Coleman 1980) and she argued
that such figures help to explain why there was a 44 per cent fall in land
occupied by factories, 20 per cent in docks, 37 per cent in railways and 38 per
cent in residential and commercial buildings in Tower Hamlets between
1964 and 1977. The other side of this question is the low cost of holding land
in a derelict or vacant state. In Britain there is no national or local tax upon
vacant land, and thus little financial penalty attaches to keeping it unused
(Chisholm and Kivell 1987; Cameron 1988).

Specific causes

The separation of the causes of dereliction and vacancy into general and specific is an arbitrary division of convenience. It simply attempts to separate out those factors which are widely pervasive at a national or international scale from those which are more localised in their impact upon particular sites or localities.

Industrial decline

The localised impact of the closure or movement of manufacturing plants has been one of the main contributors to the creation of vacant and derelict land. The problem has disproportionately affected large, old cities, and in many cases extensive tracts of wasteland have been created. Two main processes have been operating. The first is the death, or closure, of firms for economic or technological reasons. The second process is the restructuring or relocation of surviving enterprises, with strong movements taking place away from the cramped, obsolescent and poorly located sites in the inner city (Fothergill *et al.* 1983). With a large proportion of its cities in the 'older, industrial' category, (and with multiple other economic problems), it is not surprising that Britain's urban areas have suffered badly from industrial land dereliction.

In the Black Country, west of Birmingham, 72,000 industrial jobs were lost between 1971 and 1981 (Watson 1987), and factory closure and land abandonment occurred on a massive scale. By 1982, 1,570 ha of derelict land existed in the area, including many sites which were grossly damaged. Sheffield experienced similar conditions following the dramatic decline of its metal manufacturing and engineering industries. Between 1981 and 1984 the share of manufacturing in the total employment fell from one third to one quarter. In 1978 the city had 117,000 employees in manufacturing, but the following decade saw the loss of 59,000 jobs and a consequent halving of the total. According to Watts *et al.* (1989: 15) 'the main result of this contraction has been to create large areas of abandoned land and/or buildings. In 1986 there were 301 ha of vacant land in the Lower Don valley alone'. Some of this land has subsequently been reused, and it is interesting to see how the boundaries and site configurations of many of the new uses have been determined by the previous activity. For example, the modern Meadowhall Shopping Centre reflects the old Hadfield steelworks site and the configuration of the airport mirrors the now closed Tinsley steelworks. The closure of steelworks is by no means unique to Britain, and similar examples can be found in a number of cities in North America, Belgium, Germany and France.

Manchester is another city badly affected by industrial closure and by the early 1980s this had replaced slum clearance and transport blight as the main

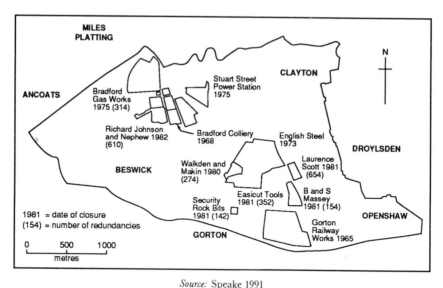

Source: Speake 1991

Figure 7.3 Major industrial closures in East Manchester

cause of wasteland (Adams *et al.* 1987). In a concentrated area of three inner city wards (Beswick and Clayton, Bradford, Miles Platting), 20,000 jobs were lost between 1971 and 1985 and one-third of industrial land and buildings lay vacant by 1987. Some indication of the spatial impact of these closures, and the resulting pattern of vacancy can be seen in Figures 7.3 and 7.4 (Speake 1991). Here, as elsewhere, the local authority has allocated much of the vacant land for future industrial use, but relatively few new activities are coming forward to redevelop such inner city sites, and manufacturing industry is rarely represented.

Dock, railway and public utility closures

Alongside the closure of manufacturing industries, the closure and changing location and space needs of a number of public utilities, railways, and above all, docks, has been important in creating urban dereliction. Perhaps because of the relative attractiveness of waterfront locations, dockland areas have often experienced relatively rapid redevelopment, albeit with substantial changes of land use as a mixture of commercial, leisure and residential activities replace old transport, distribution, processing and manufacturing functions. This transformation is, however, not always rapid, and long periods of dereliction may occur. Despite major reclamation efforts large swathes of wasteland still blight many docklands, including those of Liverpool, London and Jersey City.

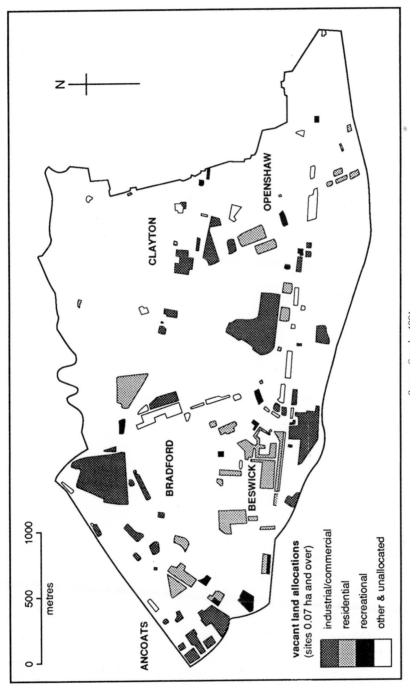

N

CLAYTON

OPENSHAW

BRADFORD

BESWICK

ANCOATS

0   500   1000
metres

vacant land allocations
(sites 0.07 ha and over)

industrial/commercial

residential

recreational

other & unallocated

Source: Speake 1991

Figure 7.4 Vacant sites in East Manchester

Most commonly, dock closures stem from technical changes such as the growing size of ships and containerisation, although there may be a host of other physical and institutional factors involved. Frequent consequences are the migration of the port to deeper water, the severing of port–city links and the creation of extensive disused waterfronts with varying redevelopment potentials (Hoyle *et al.* 1988).

The present wave of dockland dereliction, followed by redevelopment started in Boston, with the Union Wharf scheme in 1956 (Chaline 1988), from whence it spread to Baltimore, then to Canada (Toronto and Montreal), the West Coast (San Francisco, San Diego, Vancouver, Seattle) and, by the 1980s, to the Gulf of Mexico (New Orleans, Galveston, Corpus Christi). In Western Europe, it is the UK which has been worst affected by dock closures. In London, St Katherine's Dock closed in 1967 and this was followed by a 20 year programme of other closures in the capital's docks. Provincial cities, including Liverpool, Manchester/Salford, Swansea, Cardiff, Glasgow, Southampton, Belfast and many others followed. Here, redevelopment was slower than in North America, but with the return of investment confidence in the mid 1980s, a number of large schemes were successfully undertaken with government support. Notable among these have been the London Dockland redevelopment, probably the largest development scheme anywhere in Europe, Salford Quays in Manchester docks and the Albert Dock in Liverpool.

Similar processes have been operating in many other ports, including Marseilles, Brisbane and Sydney. Even such a prosperous port as Rotterdam has not been immune from the creation of derelict land through technical change (Pinder and Rosing 1988). In this case, the closure of a number of docks was accompanied by the abandonment of a water purification plant and a ship building yard at Wilton–Fijenoord.

Where dockland dereliction differs from that due to other industrial closures is in the relative success and rapidity with which new activities have been attracted. Much of the interest in waterfront development has been for housing, restaurants, retailing and recreation, essentially postindustrial activities, rather than from industry and commerce. Figure 7.5 shows an example of this in the redevelopment of Swansea docks. In London overseas capital has been an important element in the redevelopment, and land prices increased tenfold in the 1980s (Stevens 1988).

Apart from the docks, some of the largest areas of dereliction, and consequently the largest potential redevelopments have been created by railway closures. In London, the acclaimed Broadgate development and the current £3 billion scheme for the redevelopment of 54 ha at King's Cross both come into this category.

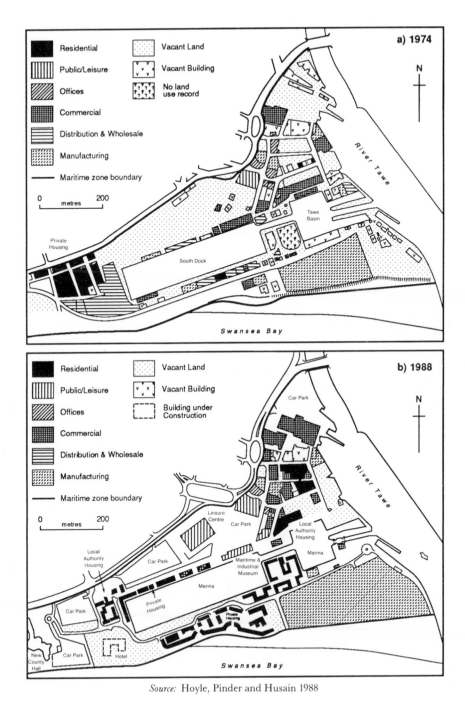

Source: Hoyle, Pinder and Husain 1988

Figure 7.5 Land use in the Swansea Maritime Zone: a) 1974, b) 1988

Minerals

In Britain, much dereliction traditionally stems from mineral workings, predominantly coal, and occurs in the form of spoil heaps, excavations, pits and subsidence. Given the close historic links between coal based industrial activity and urban growth, especially in the nineteenth century, there is a close correspondence between derelict mineral workings and urban settlements. The 1988 Derelict Land Survey revealed that 58 per cent of urban derelict land had mining origins. This factor is particularly obvious in the towns of Lancashire, Yorkshire, South Wales and the North East of England, but it is also a feature of many towns in northern France, Belgium and the Ruhr district of Germany. Even where obvious surface features, such as spoil heaps or excavations are absent, hidden mineshafts or ground subsidence may present a hazard to development.

In North America the link between urban development and mining is less close. Derelict mineral workings are common, but they tend not to be in major urban areas. Land subsidence and flooding, caused by the depressuring of underground reservoirs of water and oil, is however locally severe in a number of urban areas, including Long Beach and Santa Clara in California, and Houston–Galveston in Texas.

Physical constraints

When it comes to a consideration of specific sites, there are frequently a number of physical aspects of dereliction which impose constraints upon redevelopment. Technically, there are very few problems which cannot be overcome, but cost is often the deciding factor. Constraints may include site size and shape, access from roads, service and infrastructure provision, the image of the area and adjacent uses, ground conditions including cellars, foundations and toxic residues and previous or present planning blight. All, or any of these problems may be sufficient to deter a developer, especially in conjunction with other considerations such as poor local demand, high inner city land prices and easier or more profitable sites to be had on the urban fringe.

LAND RESTORATION POLICIES

Given the nature of the problem represented by derelict land, it is not surprising that successive British governments have, for over a quarter of a century, operated various policies to encourage its reclamation and reuse. During the latter half of this period, additional measures have also been applied to vacant land.

Initially, the problem was seen as one of dereliction inherited from mining activity, and legislation was addressed to the prevention and cure of this, especially through grant aided programmes of environmental improvement. By the end of the 1970s the derelict/vacant land problem was increasingly connected with industrial closures and urban decay. In 1979 a change of government brought a change of philosophy, away from state intervention towards a greater emphasis upon the free market. As a result of these two changes, the approach to land reclamation moved away from simple environmental improvement towards wider objectives of bringing about urban regeneration by stimulating private sector investment in such 'hard' uses for reclaimed land as housing, offices, retailing and industry. In fact, of the reclaimed land actually brought into use during the period 1982–8, 'hard' uses, i.e. industry, commerce and housing accounted for only 27 per cent of the total area (DoE 1989). Virtually all of the derelict land reclamation policy since 1980 has depended upon the government's belief that derelict land should play an integral part in urban policy, and that modest amounts of government expenditure should be used in a pump-priming fashion to generate larger sums of private sector investment. Partly this is an exercise in getting maximum value for public money. Government figures (HM Government 1984) suggested that in such schemes, the gearing ratio of public to private investment was typically 1:6, but occasionally went as high as 1:50.

A wide variety of measures relating to the reuse of derelict and vacant land have been applied in the 1980s (Kivell 1987; Kivell 1989; DoE 1989). For vacant land, the main aim has been to encourage owners, especially public sector bodies, to put their land on the market through such devices as the land register, or the Land Authority for Wales. In the former case, powers of compulsory sale may be used; in the latter case, land is routinely purchased by the land authority and then, after restoration or site amalgamation, it is sold on for development.

For derelict land, the general approach has been to make available government grants, operated by the local authorities and varying in their level across the country, in order to pay for land reclamation work or to make the process of redevelopment more financially attractive. In round figures, a budget of £650 million of public expenditure funded some 8,500 ha of reclamation in the period 1982–8 (DoE 1989). The emphasis upon 'hard' end uses, which has dominated since 1982, is undeniably desirable for regenerating urban economies, but because of the nature of inner city sites and the higher standards of ground work necessary if building is to take place, costs are much higher than where reclamation is for open space or agriculture. According to the Department of the Environment (DoE 1989) average costs of reclamation schemes undertaken with Derelict Land Grant in 1987–8 were:

For housing   £163,000/ha
For industry   £129,000/ha
For public open space   £25,000/ha

Table 7.5 Urban development corporations and derelict land

UDC	Date established	Area (ha)	Original nature	% originally derelict
Merseyside	1981	960	Disused docks and associated facilities	80
London Docks	1981	2,070	Docks and industry	45
Trafford Park	1987	1,267	Industry and transport	33
Black Country	1987	2,345	Metal working and other industry	
Teesside	1987	4,565	Steel, chemical and other industry	>50
Tyne & Wear	1987	2,375	Shipbuilding and other heavy industry	33
Cardiff	1988	1,093	Mainly dockland	25
Manchester	1988	187	Miscellaneous industry and urban development	40
Leeds	1988	540	Mixed industrial and power station	25
Sheffield	1988	900	Mainly steelworks	40
Bristol	1989	360	Mixed industry	20

Source: Miscellaneous Department of the Environment sources

Consistent with its policies of involving private developers and using land reclamation to promote urban regeneration, the government embarked upon a number of major high profile schemes. The most comprehensive of these are the dockland redevelopment programmes in London and Liverpool where urban development corporations were set up in 1981 to encourage and manage the total reconstruction of the areas concerned. Additional development corporations in Manchester, Tyne and Wear, Teesside, the West Midlands, Leeds, Sheffield and Cardiff were later announced. As Table 7.5 indicates, the development corporations were very much concerned with the problems of industrial decline and significant proportions of their areas consisted of derelict land. Such schemes can call upon funds outside of the regular derelict land allocation. These corporations were always seen as temporary creations, and there are now indications that some of the earlier ones may be wound up in the mid 1990s. This will leave major questions to be resolved about the return of resources and responsibilities to the local authorities. More speculatively, a second high profile approach has been applied through the use of Garden Festivals as development catalysts in rundown areas of Liverpool (1984), Stoke on Trent (1986), Glasgow (1988), Gateshead (1990) and Ebbw Vale (1992). By attracting several million visitors and investment from outside of the area, these events are thought to provide an effective concentration of land reclamation effort and resources, and a major stimulus to sluggish local economies. Some successes can be claimed in terms of subsequent development (e.g. Stoke-on-Trent, where 'Festival

Table 7.6 Summary of mechanisms for land reclamation in England and Wales

---

PREVENTION

Planning control: Restoration conditions to prevent dereliction from occuring upon cessation of mining activity are imposed in the field of mineral extraction.

Pollution control: This does not normally prevent derelict land from occurring, but it can be used to regulate certain industrial processes which could contaminate land.

RECLAMATION

Derelict land grant: This is widely available to public bodies and private land owners throughout the country, although at varying levels. It is intended to help with reclamation costs and to ensure that the subsequent use or development costs are no higher than they would have been on a greenfield site.

   Total expenditure 1982–8 = £628 million
   Land reclaimed   = 8,520 ha

City grant: To encourage inner city reclamation by bridging the gap between costs and value upon completion.

   Total expenditure 1982–8 = £32 million
   Land reclaimed   = 260 ha

Urban programme: Grants to support local authority programmes to tackle underlying economic, social and environmental problems.

   Total expenditure 1982–8 = £89 million
   Land reclaimed   = 2,980 ha

Urban development corporations: To promote physical, social and economic regneration of areas through land acquisition, restoration, building and provision of infrastructure.

   Total expenditure 1982–8 = £222 million
   Land reclaimed   = 810 ha

Source: Various DoE Reports. The projects to which these schemes relate vary so much that it is not possible to make direct cost comparisons

Park' now boasts a very active complex of new commercial, retail and recreational facilities), but major questionmarks surround the outcome of others (e.g. Liverpool). Existing mechanisms for land reclamation in England and Wales can be summarised fairly succinctly as in Table 7.6.

The effectiveness of reclamation policies is not always easy to judge, partly because the direction and nature of the programme has changed from time to time, and partly because the benefits of some of the more expensive and elaborate schemes will be diffused througout a wide urban area and will take a long time to be fully felt. A recent evaluation (DoE 1987) suggested that the environment and safety objectives of Derelict Land Grant schemes were successfully met in virtually all cases and had produced significant spin-off benefits in encouraging investment confidence in many areas. The objectives relating to the provision of development land however proved more difficult to confirm, or at least took longer to achieve and many reclaimed sites remained vacant in the face of low levels of commercial demand. Whilst it is clear that there have been many successes with land reclamation policies, there remain significant problems. Some of these have been documented elsewhere (Kivell 1989), but they may be summarised as: a poorly co-ordinated overall package of urban policy measures; an emphasis upon private sector development which may unintentionally favour cities with dynamic local economies, especially in the south of England, whilst failing to deal with the hard core of the problem in the north; the slow reduction in the total amount of dereliction, which fell by only 5000 ha between 1982–8 to leave 40,000 ha still derelict; the deterioration of some earlier public open space schemes whereby reclaimed derelict land is in danger of becoming derelict reclaimed land; the lack of community involvement in reclamation.

Most countries in northern Europe have planning policies and financial aid packages to deal with damaged industrial land (DoE 1989), although there is little evidence to suggest that any of them have more successful solutions to the problem, once it exists, than the UK. All involve a degree of public sector/private sector collaboration.

In France, remedial programmes focus upon sectors (e.g. coal or steel) or upon declining regions (Clout 1988), but administrative and technical approaches are fragmented. ZACs are widely used to create land ready for development and a further instrument was created in 1982 in the form of *zones de conversion*, in areas where coal, steel, textiles or ship building were in crisis. A proportion of the budget is directed at *friches industrielles*, but the total figure is small. In addition, there is a separate fund for mining areas, and 'contract plans' are drawn up between the state and the region. Those for Nord/Pas de Calais, Lorraine and Champagne/Ardennes emphasise land reclamation.

In France, as in Britain, public sector intervention has been used to stimulate private investment and to moderate the role of the free market. Although public–private co-operation is not always harmonious it generally works well. For example on the Citroën site in Paris the combined influence of the state and local authority ensured that community services, social housing and a public park were developed alongside private housing, offices and other commercial activities. France has also used large, high profile redevelopment schemes in areas of acute dereliction, e.g. on the old Ideal Standard factory site at Aulnay sous Bois, in the decayed textile quarter of Girons in St Etienne, The USINOR steelworks site at Denain and the theme park which has been developed on the 330 ha site of a closed steelworks at Hagondange near Metz.

Germany has one of the longest and most active records of land reclamation, especially in mining areas, although as with France, different historic, economic and planning conditions have resulted in less urban dereliction than in Britain. In the Ruhr, a multi-billion mark action programme for infrastructural improvement and economic regeneration was started in 1979, with 500 million marks set aside for land reclamation (Couch and Herson 1986). This programme, the Grundstuckfond, represented an unusually high level of public intervention in land whereby the public sector invested heavily through a purpose made, privately limited company owned by a consortium of public agencies. Land which is suitable, or which is reclaimed for community use, is sold to local authorities at a nominal price, but approximately half of the land involved has been used for private sector industrial use.

In the USA there is nothing comparable in terms of a purely derelict land reclamation policy, but there is a direct parallel with British practice in that land planning interests have been increasingly expected to play an important role in the urban regeneration process. In a number of cities with large land redevelopment problems, including waterfront projects in Boston and Baltimore and Pittsburgh's inner city, public sector funds from both federal and local sources were used as a catalyst. Public–private partnerships were commonly used and although they were generally successful there were inevitably some conflicts over the balance between social and commercial needs. A bigger conflict of views was revealed in the mid 1980s, when it became clear that the administration in Washington began to regard urban development grant programmes as bureaucratic, inefficient and a threat to the proper workings of the market. As a result, the federal Urban Development Action Grant programme was severely curtailed and state and city governments were left to provide financial incentives to developers.

CONCLUSION

This chapter has suggested that derelict and vacant land is a significant part of the overall land use pattern of most cities and amounts to a major problem in a number of them. It occurs for a variety of reasons, including inefficient planning and urban development, but above all it is a product of industrial decline and the restructuring of local urban economies. This can be seen both at a very general level, as a consequence of the transformation of many western cities in the early stages of a post industrial age, and at a more specific level as individual industries and transport undertakings are abandoned in particular locations. At the general level, it is clear that the last quarter of the twentieth century is witnessing a restructuring of the urban space economy, some redefinition of what locations are desirable for different activities and some reconceptualising of urban and non-urban realms. Although the emerging pattern is not yet fully clear, it is evident that the space needs of the postindustrial metropolitan area are different from those of the industrial conurbation. In the process of transformation, it is the inner city which has experienced some of the most radical and painful land use changes.

For Britain, some of these problems of transformation appear to have been particularly acute. Internationally the information on derelict and vacant land is very poor and scant, but there is sufficient evidence to suggest that Britain's cities have been more extensively affected by vacant and derelict land than those of any other nation. Particular national circumstances contributed to the creation of such extensive areas of wasteland, and local economic, market and planning conditions resulted in much of it being relatively long lived. Only since the mid 1980s has the overall total begun to fall, but there are still individual categories and individual localities where it is not falling. In accounting for Britain's relatively greater problem, it is necessary to take into account that most other nations, with which comparisons may be made, have had:

- stronger national and regional economies;
- fewer frustrated urban housing and transport schemes;
- better co-operation between public and private sectors with local authorities either being or allowed to be more enterprising;
- a more recent history of urban/industrial development.

In dealing with the problems of derelict and vacant land, especially in their older manufacturing cities, virtually all governments have acknowledged that some public sector assistance is necessary in restoring the land, underwriting the investment risk, freeing a seized up land market and getting development underway.

8

SOME CONCLUSIONS

Cities contain major elements of stability, but they are by no means static. Change and transition are normal parts of the urban condition and the urban form is, as Harvey (1985) suggested, remarkably restless. However, several things are unusual about the present phase of transition for most western cities: first, it is a very comprehensive transition, affecting many dimensions of city life; second, it is taking place with unprecedented speed; and third, it has contained many aspects of decline, which have contrasted sharply with the era of unprecedented growth delivered by 'Keynesian' state regulated capitalism from about 1945 to 1970. After the early 1970s, urban planners, administrators and politicians had to cope not simply with the challenges of change and growth management, but with the altogether more daunting task of change and decline. This chapter will argue that much of the discourse about urban change has concerned economic, social or political transition, but that there are important land use implications which have often been neglected. As has been argued in previous chapters, land use is of vital importance because it is both the container for urban activities and thus forms the physical framework of cities, and because it is one of the keys to economic, social and political power.

If we are to understand the role of land use, and the changing demands placed upon it, we need to understand something of the urban transition itself. To a large extent, the present era of urban transition has a background of crisis. Indeed the term 'urban crisis' has been used by some as a synonym for the metropolitan restructuring which has taken place over the past two decades (Gottdeiner 1986), and others have described the present as a period of 'disruptive transition' (Cadman and Payne 1990). Not all cities have been in crisis, but the indications have been widespread, especially in the older industrial cities and weaker regional economies of the UK and the USA. The ghetto riots of the 1960s, recurrent difficulties of urban financing, the energy crisis and the beginnings of more general economic problems in the 1970s, the maturing of the economic recession and large scale unemployment in the 1980s, the crisis of public services and renewed social polarisation in the 1990s have all been focal points of the urban crisis. These have sometimes been identified as inevitable problems of capitalism and it is certainly true that the response of governments has often been to attempt to solve the problems by improving the urban environment for businesses. Urban regeneration schemes, especially those concerned with land reclamation and the assembly of development sites, have been designed to lure businesses back into the city. Social consumption investment too, in such areas as subsidised housing and transport has had major implications for land use and the shape of the city.

The city, at one level can be seen as a concentration of individual people and families. Following this thread, we can see that even amongst those who have done well, economically and socially, out of the recent changes, perhaps especially amongst those who have done well, the city holds few attractions as a place to live. In the face of multiple disincentives, such as high prices, low quality housing, unreliable public transport, rising crime figures, dirty environments and a fragmented sense of community, many of those who can, flee the city as they have done for over a hundred years.

CURRENT TRENDS

The evolving land use pattern of the city is largely determined by several fairly clearly established, and easily observable, urban and industrial changes. These are widespread in North American and European cities, although they have been taking place at slightly different points in time and with varying speeds and intensities (Hay 1990). Five broad trends need to be addressed here, although, as will become evident, they are interrelated to some extent. These trends are also working in largely the same direction by creating land use patterns which result in a loosening, even an abandonment of some of the traditional urban fabric, and the development of more dispersed urban forms and fringes.

The decline of manufacturing

In most of the established industrial nations there are many signs of industrial decline at both national and local level. In Britain, for example, manufacturing attained its peak in the 1960s since when it has declined in a number of ways. Its share of both GNP and total employment fell between 1964 and 1984 from over a third to less than a quarter, and, as Chapter 7 showed, this fall resulted in the abandonment of large tracts of land. The collapse has been particularly serious in the big cities; from 1971 to 1978 there was an overall contraction of 9.9 per cent in manufacturing employment, but a fall of 15.8 per cent in the major conurbations (Hall 1985). This was followed by a further fall of 22.7 per cent in the conurbations between 1978 and 1981 (Fothergill *et al.* 1986). Not only did large numbers of businesses close, especially in the inner city, but in their efforts to remain competitive

178

many others rationalised or merged their operations and sometimes they moved to new locations on the edges of towns where more space, or cheaper space, was available. This decline of manufacturing has left some towns entirely bereft of the activity upon which they were once based, for example, ship building has entirely disappeared from Sunderland and railway engineering from Swindon. Elsewhere there have been major shrinkages, for example, Teesside is still a major centre of metal and chemical manufacturing, as it was a generation ago, but reorganisation and decline has resulted in the creation of hundreds of hectares of derelict land and the loss of tens of thousands of jobs in manufacturing. As Champion and Townsend (1990: 207) so eloquently put it, 'The older industrial parts of Britain are often the forgotten districts of the world's first industrial country.'

A particular change within manufacturing which has broken many ties with its traditional locations, and accounted for many local declines, has been an increase in mechanisation. This has resulted in smaller labour forces being required for any given activity and hence the possibility of locating in smaller towns. It has also brought about lower employment densities (typically between 20–80 jobs per hectare for modern manufacturing) and an increase in the land required per job. At the same time, capital equipment has often become less bulky and more flexible, transport has been greatly improved, many products have become much smaller and a new range of materials such as plastics have been substituted for wood and metal. In some cases all these changes have resulted in an absolute contraction of an industry, in others, simply a decline in one location and expansion in another on perhaps a regional or even a global scale. Whilst the decline of manufacturing is very obvious in some senses (e.g. in blue collar job losses and land dereliction), a slightly different interpretation is also possible (Cohen and Zysman 1987; Gershuny and Miles 1983). According to this view, manufacturing is still vital to national and urban economies, but due to structural changes, fewer jobs and less space is given over to the actual manufacturing process. To some extent, this is counteracted by there being more jobs and space taken up by design, technical, administrative and marketing functions, but the important point is that these are commonly in different locations from the actual manufacturing process.

With the advantage of hindsight, it is possible to see that many of the conditions for the deindustrialisation of cities, especially in Britain, were established in the 1930s. It is clear that a number of serious industrial weaknesses became evident but the Second World War, and the ensuing 25 year economic boom, during which Britain's industries were shielded from competition by trade agreements within its colonial sphere, delayed the decline, and hid the necessity to tackle the problems earlier.

Throughout modern history the city has been intimately connected with national economic expansion, being both generator and recipient of economic growth. When that growth changes its direction or nature, the implications for the city are profound. Thus it is that some of the most far reaching changes to the land use pattern of the city have been caused by recent shifts away from manufacturing industry. The urban–industrial nexus, which has been so powerful for two centuries, has been broken.

Suburbanisation and decentralisation

The dispersal of urban population, in the form of suburbanisation or urban decentralisation, has been one of the most important trends in the cities of the developed world over recent decades. The process was forecast by Ebenezer Howard and H.G. Wells as long ago as the turn of the century, it was measured and analysed by Colin Clark in the 1940s and 1950s, and it has been described by Peter Hall (1985: 40) as an 'absolutely regular, absolutely predictable process'. Suburbanisation is a relatively restricted term given to a part of what has become a bigger process of urban decentralisation, which in turn has been extended to the concept of counterurbanisation. These processes have been extensively documented by Champion (1989), inter alia, and need little further elaboration here.

The outward movement of population alone has considerable land use implications, most notably in terms of the decline in residential densities, the diminishing attractiveness (and price) of land in the inner city and the growing demand for land on the urban fringe. The process is however much wider than this, for it commonly includes job decentralisation, suburban employment growth (Pivo 1990) and the building of offices and shops outside of the city centre. This leads to an increasingly complex web of cross-cutting suburb-to-suburb journey patterns and the creation of dispersed urban forms in which the suburbs, or the periphery have replaced the core as the main focus of activity. Twenty years ago, Berry and Cohen (1973) pointed to the radical restructuring of metropolitan America which was being brought about by the decentralisation of commerce and industry. By 1986, over half of US office space was located outside of urban downtowns (Fulton 1986), and it was clear that office suburbanisation was a major element shaping the new metropolitan forms.

New economic activities and locations

The traditional role of the city, as the centre of economic activity, has been challenged both by the changing locational preferences shown by established activities and by the emergence of new economic activities with new land use and locational requirements.

Amongst the new activities there is a group of high technology under-takings which has become particularly important in shaping the new space economy. Covering a wide range of fields, including computer hardware and software, biotechnology, medicine and pharmaceuticals, precision instru-ments and a wide variety of scientific applications, the high-tech field has created new land use and locational patterns. In particular, the mixture of knowledge and data based activities, venture capital, research and development, government contracts and small-scale, high-value manufacturing which is involved favours high amenity areas with highly qualified workforces. The image of the old industrial city, with its degraded environment and poorly skilled, inflexible workforce is repellant and the sites favoured for high tech development are usually at some distance from the older urban–industrial centres (Hall and Markusen 1985). In this way, developments have grown up on spacious sites around small towns, university campuses and urban fringes in the newly favoured locations such as the eponymous Silicon Valley to the south east of Paolo Alto in California, along Boston's route 128, or, on a modest scale, along the M4 and M11 corridors from London to Bristol and Cambridge respectively.

High-tech industries are relatively new, and are establishing their locational and land use requirements from scratch, but there are also many other established activities which are now responding to new locational imperatives. Existing manufacturing industry, in many cases, is reacting to urban congestion and high land prices by moving out of the city, and firms which are setting up for the first time, for example, new Japanese industries in the UK, commonly select greenfield, new town or urban fringe locations. Retailing, wholesaling and distribution depots, which have long recognised the advantages of urban fringe and out of town locations in North America, have recently followed the same pattern in Europe, particularly in response to new motorway networks. Even offices, traditionally a bastion of the central business district, are increasingly striking out into new locations.

The locational processes which are at work are sometimes very obvious; for London they include congestion, the high price of land and the poor quality of public transport. Less obvious, but of growing importance for labour recruitment into the more sophisticated new activities, are such factors as inadequate schools and the shortage of particular skills (Henley Centre 1990), and the amalgam of 'lifestyle' factors which defy definition. A further set of factors which affect the locational pattern is connected with business mergers and acquisitions. In the USA, many major firms listed in *Fortune* magazine have decentralised their headquarters from the largest metropolitan areas. In the 1980s, New York continued its decline in this respect for a third decade and was joined by previously dynamic office centres such as Los Angeles and Houston (Holloway and Wheeler 1991). If the big city centres were the chief losers, the chief gainers were the suburbs and freeway corridors (Pivo 1990), the small/medium sized non-industrial cities (Hay 1990) and the growing office parks on the urban fringes. One of the main characteristics of the 1980s, both in the USA, where they represent the continuation of an established trend, and in the UK where they are relatively new, has been the growth of industrial estates, business parks and retail complexes on urban fringes and at motorway junctions.

Technological change

In a very obvious way, the form, structure and land use patterns of cities reflect levels of technological development. Indeed when such adjectives as western, developed, modern, Third World or industrial are used to describe cities, the subtext is commonly the level of technological development. In western nations, technology is changing very rapidly and it has an impact upon almost every facet of urban life. Within the field of manufacturing, which has provided the main motor of urban growth for two centuries, technology has recently altered the power sources, the manufacturing processes themselves, the transport inputs and the labour requirements; because of all this, it has also altered the locational preferences of firms. Factories are now often smaller, use more land per employee but less in total and require smaller labour forces with different combinations of skills from previously. This has given them more flexibility than before and the ability to seek cheaper locations, for example, in the suburbs or further afield.

A major contemporary development is that of information technology (IT). This has already had a profound effect upon the way in which manufacturing industries function and locate, but more than that, in a number of cities, information processing has replaced manufacturing as the chief economic activity. Above all, it is computer technology which is driving the shift in production and employment out of primary and secondary industries into the service sector (Harris 1987). Much of the high technology activity requires specialised linkages within a sophisticated economy, and a highly educated workforce. For these reasons it is likely to remain associated with the infrastructure of complex metropolitan areas, including the universities, although it is not attracted to the metropolitan cores. However, there are also parts of high-tech industries and the new information economies which require lesser skills and facilities. Routine data processing and some of the lower level assembly work, for example, could be accomplished by cheaper, unskilled labour (May 1990) located outside of the main western metropolitan economies.

Alongside the computer technology there is also a growing range of telecommunications facilities, including telephones, fax machines, fibre optic links and satellite systems which are bringing about changing patterns of employment, changing fortunes for urban locations and changing land use demands. Such links allow the distant separation of design, assembly, administrative and marketing functions in manufacturing industry and have given rise to whole new fields of activity within service industries such as finance, insurance and the mass media. Employees in many organisations have been freed, either wholly or partially, from a fixed office or institutional base. The Henley Centre forecasts that by the end of the century, almost half of all employees could telecommute for half of their work. Traditional patterns of home and workplace location, and of commuting between them

could be broken down, but there is doubt about whether the effect of telecommunications will be to increase or decrease physical travelling (Breheny
1991).

In addition to affecting land use patterns through the detail of home and
workplace location, the shift to an information economy will undoubtedly
affect existing urban hierarchies, resulting in a differential pattern of urban
growth and decline similar to that provoked by the coming of the railway in
the last century. In England, Hall (1987) has suggested that London has held
its own as a world city, but that information technology has highlighted the
way in which the ten or so major provincial cities represent the focal points of
regional systems that are archaic or are in decline.

It is not just in the sphere of work that information technology is having
an impact upon urban structure and land use. In the field of entertainment,
the video recorder, satellite TV and the home computer have changed leisure
and social patterns and have introduced a new global scale of activity. In
shopping too, a number of test schemes for the computerised ordering of
goods for home delivery from large warehouses point the way to new
patterns. PRESTEL, ORACLE and TELETEXT have all brought a small
measure of IT into British homes, and in France MINITEL has taken it a
stage further and made it available also in many schools, cafes and motorway
service stations. New York is probably the world's most connected metropolis: businesses have access to online computer databases, telephone
systems with conference and call forwarding facilites; PCs and lap top
computers are commonplace; facsimile machines are now beginning to
follow telephones into motor cars; more than sixty TV channels are available; buildings are festooned with microwave dishes and vehicles have forests
of aerials; even motorcycle messengers and pedestrians boast cellular phones
and pocket sized electronic organisers.

One of the potentially biggest impacts of IT has been to alter, and sometimes to remove, one of the main activities for which cities exist, that is the
business of people meeting together for exchange and face to face contact.

Information technology is not the only contemporary technological
change to affect the pattern of urban activity. Equally important changes
have been happening in transport. Here it is the motor car which is particularly important, and although it is by no means new technology, it was not
until the car reached mass ownership that its effects upon urban structure
were fully felt. In America since the 1930s, and in Europe since the 1950s, the
city has been increasingly structured around the needs and possibilities of
the car. Again, both local impacts, such as the relationship between home,
workplace, shops and other focal points, and wider issues such as the growth
or decline of cities according to their motorway connectivity, are important.
At an even larger spatial scale air transport has determined some patterns of
urban growth and it is possible to identify that the great hub airports of
London, Amsterdam and Frankfurt have become key nodes in the European

economy in just the same way that Hall (1991) identified this role in the USA for centres such as Chicago, Dallas, Denver and Minneapolis.

Social/lifestyle trends

Most of the factors above are economic, and it is certainly true that economic forces are the most powerful ones shaping the structure and land use of the city. It is important, however, also to consider a number of social factors because, as many of the old economic influences fade, the social choices which people can make in increasingly affluent societies become more significant.

Social factors tend to be more difficult to specify than economic ones because they are more personalised and eclectic. Nonetheless, the importance of social factors in structuring cities and their land use patterns has been recognised at least since the work of Burgess, Hoyt and others in the 1930s.

Housing choice is perhaps the most obvious way in which social preferences are translated into a land use influence, and this takes us fairly directly back to the earlier discussion of suburbanisation. It was the contention of Patrick Geddes that the essential needs of a house and family are room and more room. People like space and the present preference seems to be for private space in the form of an individual family home with garden in a broadly suburban mode. House builders recognise social characteristics by constructing appropriate houses. Much new housing is aimed at specific groups with their own land and locational requirements, e.g. low cost starter houses, exclusive family houses or estates with a rural image. Social trends affecting marriage and divorce and the needs of an elderly population are also affecting the housing market. In Britain the generation which became home owners for the first time in the 1960s is now reaching retirement. The locational choices which they make, and the capital tied up in their property will now begin to affect the urban scene.

Shorter working weeks and shorter working lives, together with greater leisure and greater affluence will all lead to new lifestyle preferences. This will influence city development both through where and how people choose to live and how they spend their free time. Already the growth of tourism and leisure has had a major impact upon urban developments, sometimes in unlikely settings. Many of the largest urban regeneration schemes in recent years have included major and successful visitor attractions in what, until recently, might have been considered impossible locations. The development industry itself is becoming increasingly packaged, producing shopping centres, festival sites, entertainment complexes and other islands of development in often discordant surroundings (Cherry 1991). Local authorities have become increasingly aware of the need to market their cities and to project attractive images in order to boost their competitive position. The organisation of urban spectacles and highly managed urban spaces to attract people and capital has been widespread. Examples in the USA include South Street Seaport in Manhattan and Harbor Place in Baltimore. In Britain, London's Dockland, Liverpool's Albert Docks, Gateshead's Metrocentre, Birmingham's Convention and Exhibition centres and Sheffield's World Student Games in 1992 all fulfil these roles.

Social and lifestyle trends are particularly susceptible to short term swings of fashion, and are therefore particularly difficult to predict. One trend which is, however, clearly set to have a significant impact upon urban form and land patterns, is the growing public concern for the environment and 'green issues'. The urban environment itself, and the impact of city growth upon the surrounding countryside forms the focus of this concern, but broader issues, such as the energy and resource implications of urban development are also beginning to arise. The twentieth-century city has been a massive consumer of resources and has generated many forms of waste and pollution. Only now are we beginning to address the notion of sustainable cities and to consider the need for a full scale environmental reappraisal of the city (Hardy 1990), especially including the role of the motor car.

TRANSITIONS

It is clear from the above that the city is in transition, and that there is a two-way relationship between the land use pattern and the built urban form on the one hand, and the economic and social activities of the citizens on the other. It is also clear that the structure and form of the city is being tested and moulded by ever shorter cycles of technical and social change. A new city is emerging in response to new demands and new forces. In the same way that Tolstoy asked, at the end of *War and Peace*, 'What force moves the nations?', we might ask 'What force moves the cities?' Many of those forces have been outlined above, but what is still far from clear is what form the urban transition is taking in response to those forces. As Gappert (1985) put it, many agree that we are now post something, but disagree on post what.

Initial examination suggests a wide range of possible transitions:

From industrial	to postindustrial
From material flows	to information flows
From modern	to postmodern
From mechanical	to electronic
From public welfare	to privatism
From compact/suburban	to spread metropolitan
From mono-centric	to poly-centric

One of the most fully debated theses is the transition from industrial to post-industrial (Drucker 1971; Bell 1974; Touraine 1974). In essence, this involves a change from the production and consumption of goods to services, white collar workers replacing blue collar workers, the rise of a technocratic elite, the growth of informational services and the subordination of the firm to social improvement. Drucker (1971) explained that while the nineteenth-century industrial city was founded on the industrial worker, the megalopolis of postindustrial society would be founded on information and the knowledge worker; a theme which was taken up by Gottmann (1983) in what he called the transactional city. The implications for urban form and land use are enormous and can already be observed. The university campuses and low density science parks are becoming the cutting edge of new settlement forms, in place of the factory chimneys and cramped urban environments of yesteryear. One corollary of this is that the investment and energy expended upon propping up the declining industrial economies may be not simply wasted, but actually harmful if it diminishes the amount available for the new opportunities.

A similar postindustrial argument has been evinced by Castells (1990). His view is that the first great modern transformation was from agriculture to industry, and that we are now in a second phase shift, from an industrial to an informational mode. Like Bell, he argued that material resources have been largely replaced by knowledge resources as the major inputs, and that this is creating a new patterning of economic space. Three kinds of localities are important; above all, are those which provide an innovative, knowledge rich, synergistic environment for the higher level activities; second, are the international command centres such as New York, London and Tokyo; and third, are the locationally flexible areas for routine office and manufacturing functions. Crucial to this transformation is the change from human/mechanical to electronic ways of doing things.

An alternative perspective, although one which owes something to the postindustrial concept, is the transition from modern to postmodern. There are different views of this but we can follow Harvey (1989) in suggesting that modernism is generally perceived as positivistic, technocratic, rationalistic, involves standardisation of knowledge and production and planning for ideal social orders. Postmodernism, on the other hand, involves heterogeneity and difference as liberating forces, and is characterised by fragmentation, pragmatism, even indeterminacy and chaos theory. Two interrelated strands have relevance here, one primarily economic and the other more concerned with urban structure, planning, land use and architecture. The economic theme is most closely linked to the postindustrial concept, indeed some see postmodernism as being largely about the run down of heavy industry (Lyotard 1984). This can also be extended into a consideration of postmodern as post-Fordist (Goodchild 1990), in which information based industries and flexible means of production dominate. The huge, corporate industrial plant, with standardised procedures and products gives way to greater variety and flexibility, including subcontracting, decentralised decision making and job enrichment policies.

This strand of postmodernity has obvious implications for urban growth, location and land use, but there are even stronger implications in the second strand. Here postmodern can be seen as a break with the kind of planning and urban development which had its heyday in British cities in the 1960s and 1970s. It rejects the large scale, comprehensive, standardised planning and architecture of factory concentration and housing estates, and it abandons the technologically rational and sophisticated plans based upon standard ratios of space, land use or population density. British planning went particularly far down this modernist path in the postwar years and unfortunately it produced many poor urban environments in the inner city and on urban fringe housing estates. Ironically, these schemes were motivated by the best of intentions and were designed to promote social welfare and to reduce social inequalities.

The postmodern approach to planning, architecture and urban development sees a more fragmented, more diverse urban fabric. Community needs, small scale developments and vernacular architecture are all key elements. The modernist period led to functional, indeed to mono-functional, land use zoning, one result of which was to necessitate wasteful journeys between home, work and shopping areas. In postmodern designs, rigid zoning is out of favour and mixed land use development with greater diversity is promoted. Krier (1987) envisaged a 'good city' scheme in which all functions are provided within compatible and pleasant walking distance. One of the most influential books on the postmodern approach to urban planning, althought the term was hardly recognised at the time, was 'Death and Life of Great American Cities' by Jane Jacobs (1961). More recently it is the work of David Harvey (1989) which has developed the concept more fully.

A related theme, which gives the urban transition a further dimension, is the shift from public welfare to privatism. In its broad context the welfare state is taken as a particular system for the distribution of important goods and services, such as health, education and defence, through public institutions to all citizens, regardless of market forces. In the 1960s in the USA (Castells 1989), and in the UK, three trends within welfarism affected urban areas. First, the share of government spending devoted to welfare increased substantially, second, funds were increasingly distributed from central government to local government and third, public funds became increasingly targeted upon specific urban problem localities. In the late 1970s, political decisions were taken, again in both countries (but with far wider ripple effects), that rising public expenditure was handicapping market forces and restricting the ability of governments to revive their flagging economies. There was also public support for tax cutting measures, exemplified by California's Proposition 13. Governments thus attempted to trim back public expenditure on welfare and to encourage market forces.

The individualism and market oriented intervention of the New Right can be linked to post-modernism (Goodchild 1990) and there is clear evidence that the main motor of urban development in the past decade has been the private sector (Cherry 1991), with the state playing only a shared role in the development of housing, industry and othe economic activity. In fact public expenditure continues at a high level. In many British inner city communities welfare cheques remain the main economic input, and the crisis of the welfare state has regional implications for those cities which are dependent upon it. Perhaps even more significantly, as Castells (1990) pointed out, the demise of the welfare state would remove the safety net for many inner city residents, provoke strong resistance from politically militant minorities and further encourage social polarisation and the development of the dual city.

All of these transitions undoubtedly have effects, qualitatively, quantitatively and locationally, upon the urban land use pattern. By far the most important effects, however, are those caused by the transition from a relatively compact city, usually with a single centre, to a dispersed one with many centres. It has already been suggested that decentralisation has been one of the most powerful urban forces of the twentieth century, but this is not simply a matter of the city growing outwards. More than twenty years ago Ash (1969) drew attention to the increasingly open form of the city. Ward (1990) has recently pointed out that, long ago, Howard, Geddes and Kropotkin were all imagining poly-nuclear city regions and Hall (1990) has revived Melvyn Weber's concept of non-place urban realms and Gottman's megalopolis. In a recent series of essays about the changing face of urban Britain (Cooke 1989), the falling apart of traditional urban/industrial centres and the the urbanisation of small towns and the countryside feature prominently.

The tendency of cities to disperse, or spread out is normally explained in terms of both pull and push factors. The pull is provided by the attractions of a suburban or pseudo rural environment together with the advantages of land price and availability. The push is provided by the crowded, expensive, unpleasant, unsafe environment of many inner city districts. Attempts to provide modern, high density developments during the 1950s and 1960s, often following a debased Le Corbusier model, resulted in many unpopular schemes which simply provided further impetus for urban decentralisation.

The spread of the city has been comprehensive; it is not simply people who have dispersed, but also jobs and many other activities. Perhaps equally important is the fact that peoples' perceptions of desirable locations have changed and the city has lost much of its prestige and urbanity. Urban–rural boundaries are increasingly blurred and the relationship between city and countryside is shifting. At the extremes the two may still look very different in land use, but functionally they are becoming more closely linked. Of course this position should not be exaggerated. In Europe, especially, large numbers of people live outside of major cities and retain distinctive cultures and lifestyles. Throughout the western world though, an increasing reality is that of the metropolitan region where a constellation of urban places is loosely grouped around a major centre, which may itself be waxing or waning.

Within these metropolitan regions the pressures for land development may be great, but the pattern is a fragmented one with no clear sense of urban expansion from a central point. Most of the land use decisions are essentially local and individual, but there is little evidence, from either North America or Europe, that there exist the institutions or the techniques to deal with the larger land use and planning needs of the spread city. In Britain, much reliance has been placed upon the green belt to regulate land within such cases, but even this can be claimed to be an anachronism because it is drawn too close to major cities to have any real effect upon growth in the outer fringes (Herington 1991).

Urban sprawl has always had its critics, and planners, in particular, have argued that it is uneconomical, wasteful and aesthetically unpleasing. In addition, there have been strong arguments in Europe against allowing cities to spread because of the threat to land used for food production. Recently the arguments for keeping cities as compact as possible have found a new focus around green issues and energy needs. For example, the European Commission's 'Green Paper on the Urban Environment' (Commission of the European Communities 1990) refers to the failure of the urban periphery, the absence of public life, the paucity of culture and the time wasted in commuting. In contrast, it advocates compact, high density cities, as efficient concentrations of people, amenities and services. At the same time it is aware that even the use of the word 'city' may be a reference to the vocabulary of the past which may impede our understanding of new settlement realities. Other groups too, including Friends of the Earth (Elkin 1991), argue for high density, sustainable, cities in which the role of the motor car can be minimised. Whilst arguments against spread cities need to be taken seriously, there are counterclaims. For example, it is important to remember that high density urban solutions have been widely unpopular in the past. Even on energy saving grounds, the compact city is an unproven advantage, although it is evident that energy may be saved by minimising the use of the car. Here European cities are probably well ahead of their American counterparts. Sample studies have shown that North American cities, with population densities only one-third the European level, have per capita gasoline consumption rates which are four times as high (Newman 1990). However, if renewable energy sources are to be exploited fully, there are considerable arguments in favour of dispersed cities (Ward 1990). Above all, most renewable forms of energy, including solar energy, wind and tidal power can best be utilised in small packets in a dispersed settlement form.

Leaving these arguments aside, it is clear that the reality of recent years has been for the city to spread out, and the strength of this shift should not be underestimated. Pacione (1990) described the centrifugal movement as the predominant socio-spatial trend in advanced capitalist countries in postwar years. In the 1980s, British cities began to follow where American cities had led for several decades. This was due partly to a new spirit of market led development in which ample investment funds were available (at least towards the end of the decade), partly to the relaxation of planning constraints and partly to the determination of local authorities to gain what they saw as their due share of new investment. Notwithstanding some of the inner city renewal efforts, it was on the urban fringes that the major changes were felt. Some of these changes have been documented for earlier time periods (Johnson 1974; Bryant *et al.* 1982; Herington 1984), but by the late 1980s a new and aggressive phase of development pressure was being experienced and planners were under severe pressure to relax their opposition to urban development on greenfield sites.

A series of marine metaphors has been popular in describing the expansion of the urban fringe. Boyce (1966) referred to 'wave like' growth, Steeley (1990) described 'urban shorelines' in which urban life will evolve, and Hart (1991) described the 'perimetropolitan bow wave' cutting through a jumble of contradictions and mixed land uses in the urban–rural fringe. Here change is the rule, not the exception, and without strong planning intervention, urban expansion invariably wins because the farmer cannot afford to pay urban prices for land. Hart further argued that the main value of agricultural land on the urban fringe is not for food production, that can be done elsewhere, but for providing the sense of wellbeing that people derive from open space.

In this urban fringe, it is the growth of family housing which takes most land, but it is often the gleaming new, single storey factories, warehouses and office buildings which make the greatest impact. The characteristic feature of a motorway approach to many cities today is an archipelago of big sheds. Commercial developments have been particularly attracted to urban fringe locations, especially where motorway access is easy. Town and city centres are facing increasing competition from out of town shopping centres such as Meadowhall (124,000 square metres) in Sheffield or Thurrock Lakeside (also 124,000 square metres) in Essex. Business Parks too are flourishing, as at Aztec West on the northern fringe of Bristol adjoining the M4/M5 motorway junction. This development employs 4,000 people in a mixture of high-tech manufacturing and office activities, located in good quality, modern buildings in an attractively landscaped site. Twenty-three of the sixty-three firms are concerned with computing or communications, and they include IBM, ICL, Wang, and Texas Instruments. Hewlett Packard also have a large establishment less than a mile away. Aztec West is the latest addition to a complex of urban fringe developments, adjacent to the M4/M5 interchange, which include two earlier industrial/trading estates at Patchway and Bradley Stoke, and a major retail park at Cribbs Causeway (Figure 8.1). Another development, which neatly illustrates a number of relevant themes, is to be found at Stockley Park in West London, which when completed will be Britain's largest Business Park with 232,000 square metres of floorspace. The joint British/Japanese developers claim an ideal marriage of environment and location, being next door to three motorways, two railway lines and one of the world's biggest airports. It offers a range of handsome office buildings set within a landscaped parkland setting of lakes, gardens and wooded plantations. The site, which was previously 162 ha of rubbish filled gravel pit, was in a green belt, but this condition was relaxed in exchange for an estimated £60 million worth of planning gain, in the form of new roads and leisure facilities. In 1990, there were 800 business parks in the development pipeline in Britain. Between them they involved 26 million square metres of floorspace, but only a fraction of this will actually get built in the foreseeable future. In the decade 1980–90, land equivalent in area to twice the size of Bristol was earmarked for business park development throughout Britain.

One major problem with developments like Stockley Park, which is only now being realised, is that they are almost totally dependent upon the motor car. They are simply too far away from any other developments, or from recognised central areas, to make walking journeys feasible. It is largely for this reason that a newer generation of business parks, such as that at Chiswick in West London, are being designed to fit in with public transport networks.

The cumulative effect of many of these changes, within the context of urban spread, has been to turn the economic life of the city inside out and to place increasing demands for land from a range of activities on the fringe. At the same time, the whole notion of the city is becoming more fragmented and dispersed. Peter Hall's view (1990) is that the regional metropolis is the new reality. The average person rarely visits the centre, and in any case the centre is losing its monopoly as the main business life of the city can now be carried out elsewhere. Within the outer metropolitan zone, people commute relatively short distances to suburban schools, offices or institutions and they have a choice of spatial identities.

In America, Garreau (1988: 51) concluded that the urban future has already been declared in what he calls 'Edge Cities – high rised, semi-autonomous, job-laden, road clogged communities of enormous size, springing up on the edges of old urban fabric where nothing existed ten years ago but residential suburbs or cow pastures'. He sees them as the biggest change in the way Americans live and work for a century. The metropolitan area is growing, not from the centre but by spawning new Edge Cities on its fringes. There are fourteen Edge Cities in the Washington DC area alone, including Tysons, one of the best known in America with more office space and white collar jobs than downtown Miami. By these measures each of Washington's Edge Cities is, or soon will be, larger than the Virginia state capital of Richmond. Many other Edge Cities exist, including Irvine, California, the Galleria area west of Houston, King of Prussia near Philadelphia and Phoenix, which is practically made up of Edge Cities.

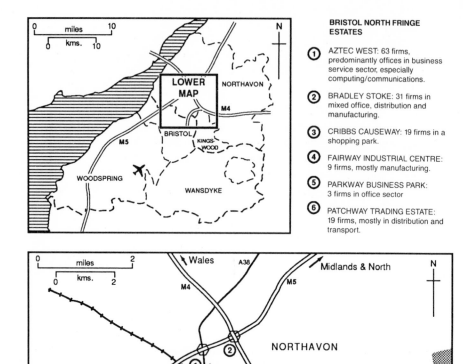

Figure 8.1 Recent developments on the northern fringe of Bristol

Table 8.1 Components of change in the city

City centre: Changing rapidly in detail, but largely retaining dense mix of retail, office and leisure uses. All of these are experiencing competition from new out of centre retail and business parks and Edge Cities. Land prices and land use competition remain high if the local economy is flourishing but a poor local economy will be reflected in high levels of vacancy.

Inner city: Changing profoundly as it tries to cope with the loss of its role for manufacturing, public utilities and transport, and a continuing decline in its status. It was the dynamic fringe area of the city in the expansionary period of the nineteenth century but is now often an area of widespread dereliction and abandonment. Restructuring and redevelopment is producing islands of investment, surrounded by a sea of decline.

Suburbs: The residential suburbs are generally the most extensive land use type within the narrowly defined city and represent a zone of stability. Detailed changes take place through infilling, and conversion of family houses, but generally there is little land use or structural change. Some commercial expansion takes place at selected nodal points.

Urban fringe: The area of maximum land use change. Change takes place from rural to urban, but what urban? Some takes the form of uniform swathes of housing, but much is a mixture of commercial/retail/business parks utilities, institutions and road networks. The fringe is increasingly the focus of activity for the postindustrial city.

FUTURE URBAN CHANGE

The nature and structure of the city are both changing, in some cases with great rapidity. In consequence, the land use pattern is also changing, but land and buildings are relatively fixed features which change more slowly than the social and economic forces around them. For this reason, land use patterns often lag behind land use processes and in extreme cases there may be a time gap of several decades between them. Different parts of the city are changing in different ways, and much depends upon the state of the local and national economies. The broad patterns of change can be summarised as in Table 8.1, but it should also be noted that some of the changes are so profound as to render this typology of urban areas inappropriate for the emerging, spread metropolitan regions.

In earlier chapters we have seen some of the processes which are at work, and the effects which they have upon urban land use patterns. Finally, at this stage it might be appropriate to look briefly at the way in which the pattern is likely to evolve.

Despite the rapidity of some of the changes of the past decade, cities can be seen as having many stabilising influences, above all in their patterns of land use, ownership and planning. Overnight transformations are not therefore likely. Cities evolve from their historic contexts, and in Europe especially, this dimension is often carefully guarded. At any point in time cities contain many remnant land use patterns and so they reflect past, as well as present, needs. Almost certainly the city a generation from now, or even two generations from now, is likely to be very similar to the city of today, although within this framework the daily way of life may be very different.

In most of the western world, the immediate historic context is one of dispersal and decentralisation. Despite small signs of renewed population growth in recently declining metropolitan cores (e.g. London), some very strong forces of dispersal are established. Given the choice, people seem determined to buy as much private space as possible, and that space is most available on, and beyond, the urban fringe. Similarly, in seeking the land, locations and images which they require, the majority of new commercial activities appear to be attracted to the urban outskirts. If, as seems clear, the structure of the city is determined by a combination of economic forces such as locational advantage, land prices and transport costs, together with social forces such as lifestyle, residential mobility, privacy and status, then an extension of growth on the major urban fringes and in attractive small town locations looks set to continue.

An unlimited continuation of these trends is, however, by no means assured. In particular, the cost and availability of energy, and the impact which this will have upon patterns of movement cannot be predicted with accuracy. Another major uncertainty is the role which planning will play. To some extent planning is an institutionalisation of underlying economic and social forces, but it is the balance between their promotion and control which is crucial. In the 1980s, planning in many western cities was diverted from its earlier, high minded concern with the creation of ideal environments, the imposition of physical order and homogeneity of land uses to promote social order, into a new pragmatism. It became more opportunistic, market led and development orientated. This is not, of itself, a wholly bad thing, but there is a negative side if the supposed advantages of development do not trickle down socially or spatially, and the distribution of benefits is grossly uneven. Although the market led approach to planning has achieved many successes in recent years, it can be argued that it does not necessarily lead to the best decisions. In particular there is the danger that it can reinforce existing biases and polarisations and neglect the need for investment in necessary public services such as education and transport.

Land use and development is inextricably linked with the social and economic forces that shape everyday life in the city. Some of these forces may appear to have a powerful life of their own, through such notions as the market or aggregated individual preferences, but through the kind of planning we chose, we can control and direct the forces. That we should do so is important, because land use and development remains one of the keys both to satisfying individual lifestyles and to the successful functioning of urban areas.

REFERENCES

1 INTRODUCTION

Beale, C.L. (1975) 'The revival of population growth in non-metropolitan America', Economic Research Service, US Department of Agriculture, ERS 605.

Berry, B.J.L. (1976) 'The counterurbanization process: urban America since 1970', in B.J.L. Berry *Urbanization and Counterurbanization*, Beverley Hills: Sage, 17–30.

Blowers, A. (1980) *The Limits of Power: The Politics of Local Planning Policy*, Oxford: Pergamon.

Brotchie, J., Newton, P., Hall, P. and Nijkamp, P. (1985) *The Future of Urban Form*, London: Croom Helm.

Carter, H. (1983) *An Introduction to Historical Geography*, London: Arnold.

Champion, A.G. (ed.) (1989) *Counterurbanization*, London: Arnold.

Champion, A.G. and Townsend, P. (1990) *Contemporary Britain: A Geographical Perspective*, London: Arnold.

Cherry, G. (1982) *The Politics of Town Planning*, London: Longman.

Cullen, J. (1990) 'Mad Maurice III, metropolitans in the making', in D. Cadman and G. Payne (eds) *The Living City*, London: Routledge.

Department of the Environment (1988) *Urban Land Markets in the United Kingdom*, London: HMSO.

Department of Transport (1986) *Transport Statistics GB 1975–85*, London: HMSO.

ESRC (1989) 'Urban and regional change in the 1980s', London, Economic and Social Research Council.

Foucault, M. (1984) 'Space, knowledge and power', in P. Rainbow (ed.) *The Foucault Reader*, Harmondsworth: Penguin.

Frey, W.H. (1989) 'United States; counterurbanization and metropolis depopulation', in A.G. Champion (ed.) *Counterurbanization*, London: Arnold, 34–61.

Friedland, R. (1982) *Power and Crisis in the City*, London: Macmillan.

Goheen, P. (1970) 'Victorian Toronto 1850–1900', University of Chicago, Department of Geography Research Paper no. 127.

Hall, P. (1980) *Great Planning Disasters*, London: Weidenfeld & Nicolson.

——— (1988) 'The industrial revolution in reverse', *The Planner*, 74, 1: 15–19.

Hall, P. and Hay, D. (1980) *Growth Trends in the European Urban System*, London: Heinemann.

International Road Federation, (1989) *World Road Statistics 1984–88*, Geneva.

Kontuly, T. and Vogelsang R. (1989) 'Federal Republic of West Germany', in A.G. Champion (ed.) *Counterurbanization*, London: Arnold, 141–61.

New Towns Committee (1946) Final Report, Cmd 6876, London: HMSO.

Stanislawski, D. (1946) 'The origin and spread of the grid pattern town', *Geographical Review*, 36: 105–20.

Vining, D.R. and Strauss, A. (1977) 'A demonstration that the current deconcentration of population in the United States is a clean break with the past, *Environment and Planning*, A, 9: 751–8.

Winchester, H.P.M. and Ogden, P.E. (1989) 'France: decentralization and deconcentration in the wake of late urbanization', in A.G. Champion (ed.) *Counterurbanization*, London: Arnold, 162–86.

2 URBAN LAND ALLOCATION

Alonso, W. (1964) *Location and Land Use*, Cambridge USA: Harvard University Press.

Anas, A. (1986) 'From physical to economic models: the Lowry framework revisited', in B. Hutchinson and M. Batty (eds) *Advances in Urban Systems Modelling*, Amsterdam: Elsevier, 163–72.

Bajic, V. (1983) 'The effects of a new subway line on Housing Prices in Metropolitan Toronto', *Urban Studies*, 23, 2: 105–17.

Balchin, P.N. and Kieve, J.L. (1977) *Urban Land Economics*, London: Macmillan.

Ball, M. (1985) 'The urban rent question', *Environment and Planning*, A, 17: 503–25.

Beckman, M.J. (1957) *On the Distribution of Rent and Residential Density in Cities*, Seminar Paper, Mathematical Applications in the Social Sciences: Yale University.

Berechman, J. and Gordon, P. (1986) 'Linked models of land use – transport interactions', in B. Hutchinson and M. Batty (eds.) *Advances in Urban Systems Modelling*, Amsterdam: Elsevier, 109–31.

Berry, B.J.L. (1967) *The Geography of Market Centres and Retail Distribution*, Englewood Cliffs, New Jersey: Prentice Hall.

Blowers, A. (1986) 'Town planning, paradoxes and prospects', *The Planner*, 72, 4: 11–18.

Burgess, E.W. (1925) 'The growth of the city: an introduction to a Research project', in R.E. Park and E.W. Burgess, (eds) *The City*, Chicago.

Capozza, D.R. and Helsley, R.W. (1989) 'The fundamentals of land prices and urban growth', *Jnl. of Urban Economics* 26, 295–306.

Carter, H. (1976) *The Study of Urban Geography*, London: Edward Arnold.

Chapin, F.S. (1965) *Urban Land Use Planning*, Urbana: University of Illinois Press.

Davies, R.L. (1972) 'Structural models of retail distribution: analogies with settlement and urban land use theories', *Trans. Inst. Br. Geog.*, 57: 59–82.

Department of the Environment (1989) *Planning Control in Western Europe*, London: HMSO.

Drewett, J.R. (1973) 'The developer's decision process: land values and the suburban land market', in P. Hall *et al.* (eds) *The Containment of Urban England*, vol. II, 163–245, London: George Allen & Unwin.

Edel, M. (1976) 'Marx's theory of rent: urban applications', *Kapitalstate*, 4–5: 100–124 also in *Housing and Class in Britain*, London: CSE Books.

Evans, A. (1983) 'The determination of the price of land', *Urban Studies* 20: 119–29.

Evans, A.W. and Beed, C. (1986) 'Transport costs and urban property values in the 1970s', *Urban Studies*, 23, 2: 105–17.

Fischel, W.A. (1985) *The Economics of Zoning Laws: A Property Rights Approach to American Land Use Controls*, Baltimore: Johns Hopkins University Press.

Fujita, M. (1989) *Urban Economic Theory, Land Use and City Size*, Cambridge: Cambridge University Press.

Garner, B.J. (1966) *The Internal Structure of Shopping Centres*, Studies in Geography, no.

Garrett, M.A. (1987) *Land Use Regulation*, New York: Praeger.

12, Evanston: Northwestern University.

Getis, A. and Ishimizu, T. (1986) 'The effect of energy costs on land use patterns in the

Nagoya Metropolitan Region', *Geog. Rev. Japan*, 59, 2: 154–62.

Goldberg, M. and Chinloy, P. (1984) *Urban Land Economics*, New York: John Wiley.

Goodchild, R. and Munton, R. (1985) *Development and the Landowner*, London: George Allen & Unwin.

Hallett, G. (1978) *Urban Land Economics*, London: Macmillan.

Harris, B. (1985) 'Synthetic geography and the nature of our understanding of cities', *Environment and Planning*, 17A: 443–64.

Harris, C.D. and Ullman, E.L. (1945) 'The nature of cities', *Annals American Academy of Politics and Science*, 242: 7–17; also in H.M. Mayer and F. Kohn (eds) 1959, *Readings in Urban Geography*, Chicago: Chicago University Press.

Harrison, A.J. (1977) *Economics and Land Use Planning*, London: Croom Helm.

Harvey, D. (1973) *Social Justice and the City*, London: Edward Arnold.

Harvey, J. (1987) *Urban Land Economics*, Basingstoke: MacMillan.

Healey, P., McNamara, P., Elson, M. and Doak, A. (1988) *Land Use Planning and the Mediation of Urban Change*, Cambridge: Cambridge University Press.

Heikkila, E., Gordon, P., Kim, J.I., Peiser, R.B., Richardson, H.W. and Dale-Johnson, D. (1989) 'What happened to the CBD distance gradient: land values in a polycentric city', *Environment and Planning*, A, 21: 221–32.

Herbert, D.J. and Stevens, B.H. (1960) 'A model for the distribution of residential activity in urban areas', *Jnl. Regional Science*, 2: 21–6.

Hoyt, H. (1939) *The Structure and Growth of Residential Neighbourhoods in American Cities*, Washington DC: Federal Housing Administration.

Hudson, R. and Rhind, D. (1980) *Land Use*, London: Methuen.

Hurd, R.M. (1903) *Principles of City Land Values*, New York: Record and Guide.

Isard, W. (1956) *Location and Space Economy*, Cambridge USA: MIT Press.

Lawless, P. and Ramsden, P. (1990) 'Land use planning and the inner cities – the case of the Lower Don Valley, Sheffield', *Local Government Studies*, 16, 1: 33–47.

Lowry, I.S. (1964) 'A model of metropolis, RM–4035–RC', Santa Monica, California: Rand Corporation.

McDonald, J. (1984) 'Changing patterns of land use in a decentralizing metropolis', *Papers of the Regional Science Association*, 54: 59–70.

Mann, P.H. (1965) *An Approach to Urban Sociology*, London: Routledge & Kegan Paul.

Massey, D. and Catalano, A. (1978) *Capital and Land: Landownership by Capital in Great Britain*, London: Edward Arnold.

Mills, E.S. (1972) *Studies in the Structure of the Urban Economy*, Baltimore: Johns Hopkins University Press.

Miyao, T. (1981) *Dynamic Analyses of the Urban Economy*, New York: Academic Press.

Muth, R.F. (1969) *Cities and Housing*, University of Chicago Press.

Neuburger, H.L.T. and Nichol, B.M. (1976) *The Recent Course of Land and Property Prices and the Factors Underlying It*, Department of the Environment Research Report no. 4, London: HMSO.

Neutze, G.M. (1973) *The Price of Land and Land Use Planning: Policy Instruments in the Urban Land Market*, 25: 325–45.

—— (1987) 'The supply of land for a particular use', *Urban Studies*, 24, 5: 379–88.

Pearce, B.J., Curry, N.R. and Goodchild, R.N. (1978) *Land, Planning and the Market*, University of Cambridge Department of Land Economy, Occasional Paper no. 9.

Perin, C. (1977) *Everything in its Place, Social Order and Land Use in America*, Princeton, New Jersey: Princeton University Press.

Pickett, M.W. and Perrett, K.E. (1984) *The Effect of the Tyne and Wear Metro on Residential Property Values*, Crowthorne: Transport and Road Research Laboratory, Suppl. Report 825.

Popper, F.J. (1978) 'What's the hidden factor in land use regulation?' *Urban Land* 37, 11: 4–7.

—— (1988) 'Understanding American land use regulation since 1970', *Jnl. American Planning Association*, 54, 3: 291–301.

Putman, S.H. (1983) *Integrated Urban Models, Policy Analyses of Transport and Land Use*, London: Pion.

Ratcliffe, J. (1976) *Land Policy*, London: Hutchinson.

Ratzka, A.D. (1980) *Sixty Years of Municipal Leasehold in Stockholm*, Swedish Council for Building Research, Document D2, Stockholm.

Rose, L.A. (1989) 'Urban land supply: natural and contrived restrictions', *Journal of Urban Economics*, 25: 325–45.

Scott, P. (1970) *Geography and Retailing*, London: Hutchinson.

Thrall, G.I. (1987) *Land Use and Urban Form*, New York: Methuen.

Titman, S. (1985) 'Urban land prices under uncertainty', *American Economic Review*, 75, 3: 505–14.

Von Thunen, J.H. (1826) *Der Isolierte Staat in Beziehung auf Landwirtschaft und National ekonomie*, Hamburg; English translation, C.M. Wartenburg (1966) *Von Thunen's Isolated State*, ed. P.G. Hall, Oxford: Pergamon.

Wacher, T. (1971) 'Public transport and land use, a strategy for London', *Chartered Surveyor*, 16.

Wendt, P.F. (1957) 'Theory of urban land values', *Land Economics*, August.

Wijers, P. (1988) 'Land prices in Tokyo', *Nederlandse Geografische Studies* 73, Amsterdam.

Williams, R. (1988) 'West Germany', in G. Hallett (ed.) *Land and Housing Policies in Europe and the USA*, London, Routledge, 17–48.

Wingo, L. (1961) *Transportation and Urban Land Use*, Washington DC: Resources for the Future.

Wiltshaw, D.G. (1985) 'The supply of land', *Urban Studies* 22: 49–56.

3 MEASURING AND MONITORING URBAN LAND

Adams, C.D., Baum, A.E. and McGregor, B.D. (1988) 'The availability of land for inner city development: a case study of inner Manchester', *Urban Studies* 25: 62–76.

Anderson, J.R. (1976) *A Land Use and Land Cover Classification System for Use with Remote Sensor Data*, US Geological Survey Professional Paper 964, Washington DC.

Berlin, G.L. (1971) *Application of Aerial Photographs and Remote Sensing Imagery to Urban Research and Study*, Exchange Bibliography no. 222, Monticello Illinois: Council of Planning Libraries.

Best, R.H. (1981) *Land Use and Living Space*, London: Methuen.

Best, R.H. and Anderson, M. (1984) 'Land use structure and change in Britain 1971–81', *The Planner* 70, 11: 21–4.

Bourke, A. and Davies, J. (1988) 'The use of GIS in Greater Manchester', *Mapping Awareness* 2, 5: 27–30.

Briggs, D. and Mounsey, H. (1989) 'Integrating land resource data into a European geographical information system: practicalities and prospects', *Applied Geography* 9, 1: 5–20.

Brown, I.D. (1985) 'Land potential and development monitoring systems', in J.R. England, K.I. Hudson, R.J. Masters, K.S. Powell and J.D. Shortridge (eds) *Information Systems for Policy Planning in Local Government*, 226–37, Harlow: Longman.

Bruton, M. and Gore, A. (1980) *Vacant Urban Land in South Wales*, Cardiff, University of Wales Institute of Science and Technology, Department of Town Planning.

—— (1981) 'Vacant urban land', The Planner 67: 34–5.

Cameron, G.C., Monk, S. and Pearce, B.J. (1988) *Vacant Urban Land: a Literature Review*, London: Department of the Environment.

REFERENCES

Central Statistical Office (1990) *Guide to Official Statistics*, 6, London: HMSO.
Champion, A.G. (1972) *Variation in Urban Densities Between Towns of England and Wales*, University of Oxford School of Geography Research Paper 1.
—————— (1974) *An Estimate of the Changing Extent and Distribution of Urban Land in England and Wales 1950–1970*, London: Centre for Environmental Studies, Research Paper 10.
Champion, A.G. and Markowski, S. (1985) *Land Use Information Held by Local Authorities, State of the Art Review: Indirect Data Sources II, National Land Use Stock Survey*, London: Roger Tym and Partners.
Charlton, M. and Openshaw, S. (1986) 'Planning and land use in a UK metropolitan county', in P.T. Kivell and J.T. Coppock (eds) *Geography, Planning and Policy Making*, 113–40, Norwich: Geo Books.
Chisholm, M. and Kivell, P.T. (1987) *Inner City Wasteland*, Hobart Paper 108, London: Institute of Economic Affairs.
Coleman, A. (1980) 'Land use survey today and tomorrow', in E.H. Brown (ed.) *Geography Yesterday and Tomorrow*, 216–28, Oxford: Oxford University Press.
Collins, M. and Barnsley, M. (1988) *Energy Saving Through Landscape Planning: A Study of the Urban Fringe*, Croydon: Property Services Agency.
Collins, W.G. and Bush, P.W. (1974) 'The application of aerial photography to surveys of derelict land in the UK, in E.C. Barratt and L.F. Curtis (eds) *Environmental Remote Sensing*, 167–81, London: Arnold.
Coppock, J.T. (1978) 'Land use', in W.F. Maunder (ed.) *Reviews of UK Statistical Sources*, VIII, 1–101, Oxford: Pergamon.
Dale, P.F. (1991) 'Land and property information systems', in M.J. Healey, (ed.) *Economic Activity and Land Use*, Harlow: Longman.
Dawson, B.R.P. (1991) 'The identification and monitoring of derelict and despoiled urban land on remotely sensed imagery: an application of texture measures', unpublished Ph.D. thesis, University of Keele.
Deane, G. (1986) 'Statistics measure post-war change', *Town and County Planning* 55, 2: 346–7.
Department of the Environment (1974) *Survey of Derelict Land in England*, London: HMSO.
—————— (1978) *Developed Areas 1969, a Survey of England and Wales*, London: HMSO.
—————— (1982) *Survey of Derelict Land in England*, London: HMSO.
—————— (1986) *Land Use Change in England*, Statistical Bulletin 86/1, London: Government Statistical Service.
—————— (1987) *Land Use Change in England*, Statistical Bulletin 87(7), London: Government Statistical Service.
—————— (1988a) *Land Use Change in England*, Statistical Bulletin (88)5, London: Government Statistical Service.
—————— (1988b) *Urban Land Markets in the UK*, London: Government Statistical Service.
—————— (1988c) *Land for Housing*, Planning Policy Guidance 3: London: Government Statistical Service.
—————— (1989a) *Land Use Change in England*, Statistical Bulletin (89)5, London: Government Statistical Service.
—————— (1989b) *Land for Housing*, Progress Report 1988, London.
—————— (1991) *Survey of Derelict Land in England 1988*, vol. 1, Main Report, London: HMSO.
Dickenson, G.C. and Shaw, M.G. (1977) *Monitoring Land Use Changes*, TP20, London. Centre for Environmental Studies Planning Research Applications Group.
—————— (1982) Land use in Leeds 1957–76: two decades of change in a British city, *Environment and Planning*, A14: 343–58.

Dueker, K.J. and Horton, F.E. (1971) 'Urban change detection systems', in *Proceedings of Seventh International Symposium on Remote Sensing of the Environment*, 1523–36, University of Michigan: Ann Arbor.

Dunn, R., Harrison, A.R. and Turton, P.J. (1991) 'Rural-to-urban land use change: approaches to monitoring and planning using GIS', *Mapping Awareness*, 5, 4: 26–9.

Fordham, R.C. (1974) *Measurements of Urban Land Use*, University of Cambridge, Department of Land Economy, Occasional Paper no.1.

Gault, I. and Davis, S. (1988) 'The potential for GIS in a large urban authority: The Birmingham City Council GIS pilot', *Mapping Awareness* 2, 5: 38–41.

Gebbett, L.F. (1978) 'Town and county planning', in W.F. Maunders (ed.) *Reviews of UK Statistical Sources*, VIII, 103–219, Oxford: Pergamon.

Gierman, D.M. and MacDonald, C.L. (1982) 'Land use monitoring in the urban-centred regions of Canada and the Canadian land data system', *Computers Environment and Urban Systems* 7, 4: 275–82.

Godfree, S. (1988) *Land Use Gazetteer*, Deal: Leaf Coppin.

Grimshaw, D.J. (1985) 'Planning application systems', in J.R. England, K.I. Hudson, R.J. Masters, K.S. Powell, and J.D. Shortridge (eds) *Information Systems for Policy Planning in Local Government*, 215–25, Harlow: Longman.

—— (1988) 'Monitoring the use of land and property information systems', *International Journal of Information Management*, 8: 188–202.

Guerin, G. and Mouillart, M. (1983) 'L'occupation des sols urbanises, *L'Espace Geographique* 2: 153–7.

Harrison, A.R. and Richards, T.S. (1987) *An Evaluation of General Purpose Classification Techniques for the Discrimination of Urban Land Use in SPOT Panchromatic and Multispectral Data*, University of Bristol, Remote Sensing Unit, Final report of MPSI Systems Ltd, Bristol.

Hathout, S. (1988) 'Land use change analysis and prediction of the suburban corridor of Winnipeg, Manitoba', *Journal of Environmental Management* 27: 325–35.

Hill, R.D. (1984) 'Land use change', *Geoforum* 13, 3: 457–61.

Home, K. (1984) 'Information systems for development land monitoring', *Cities* 1, 6: 557–63.

Howes, C.K. (1980) *Value Maps: Aspects of Land and Property Values*, Norwich: Geo Books.

Humphries, A.M. (1985) 'Property information systems', in J.R. England, K.I. Hudson, R.J. Master, K.S. Powell, and J.D. Shortridge (eds) *Information Systems for policy planning in local government*, 196–214, Harlow: Longman.

Hunting Surveys and Consultants Ltd (1986), *Monitoring Landscape Change*, vol. 2, Appendix D, Bristol, DoE and Countryside Commission, 120–30.

Jones, G. (1986) 'A summary of different land information systems', in P. Selman (ed.) *Environmental Conservation and Development*, University of Glasgow, Planning Exchange Occasional Paper 24: 160–6.

Kivell, P.T. and McKay, I. (1988) 'Public ownership of urban land', *Transactions of the Institute of British Geographers* 13: 165–78.

Kivell, P.T., Parsons, A.J. and Dawson, B.R.P. (1989) 'Monitoring derelict urban land: a review of problems and potentials of remote sensing techniques', *Land Degradation and Rehabilitation* 1: 5–21.

Levine, N. (1990) 'Demographic data for microcomputers', *Journal of American Planning Association*, 56, 4: 516–7.

Lindgren, D. (1985) *Land Use Planning and Remote Sensing*, Dordrect: Martinus Nijhoff.

McKenzie, A. (1983) 'Land for housing', *Town and Country Planning*, 52, 3: 68.

Markowski, S. (1982) *Land and Building Use: Analysis and Surveys*, Edinburgh: SSRC Planning Review, 4.

National Environmental Research Council (1978) *Land Use Mapping by Local Authorities in Britain*, Experimental Cartography Unit, London: Architectural Press.

National Remote Sensing Centre (1989) *Monitoring Change at the Urban–Rural Boundary From LANDSAT TM and SPOT HRV Imagery*, SP (89) WP56, Farnborough: Royal Aerospace Establishment.

NLUC (1975) *National Land Use Classification*, Joint Local Authority, Local Authorities Management Service and Computer Committee, Scottish Development Department and Department of the Environment Study Team Report, London: HMSO.

Norton-Taylor, R. (1982) *Whose Land Is It Anyway?* Wellingborough: Turnstone Press.

Rhind, D. and Hudson, R. (1980) *Land Use*, London: Methuen.

Sellwood, R. (1987) 'Statistics of changes in land use; a new series', *Statistical News* 79: 11–16.

Ward, R.M. (1983) 'Land use mapping techniques for city and regional planning', *Journal of Environmental Management* 17, 4: 325–33.

Weights, B. (1988) 'Kingston's GIS pilot set for promotion', *Mapping Awareness* 3, 2: 8–11.

Whitehouse, S. (1989) 'A multistage land use classification of an urban environment using high resolution multispectral satellite data', unpublished Ph.D. thesis, University of Keele.

4 PATTERNS AND CHANGES OF LAND USE

Best, R.H. (1981) *Land Use and Living Space*, London: Methuen.

Best, R.H. and Anderson, M. (1984) 'Land use structure and change in Britain 1971–1981', *The Planner* 70, 11: 21–4.

Bibby, P.R. and Shepherd, J.W. (1990) *Rates of Urbanization in England 1981–2000*, London: Department of the Environment, HMSO.

Bourne, L.S. (1976) 'Urban structure and land use decisions', *Annals of Association of American Geographers* 66, 4: 531–47.

Brisbane, W. (1985) 'Land supply for housing in urban areas', *Chartered Building Societies Institute Journal*, 39, 172: 172–3.

Brindley, T., Rydin, Y. and Stoker, G. (1989) *Remaking Planning*, London: Unwin Hyman.

British Airports Authority (1990) 'Heathrow Airport facts and figures', London.

Burtenshaw, D. and Moon, G.M. (1985), 'La Villette: problems of planning land-use change in Paris, *Geography*, 70, 4: 356–9.

Champion, A.G. (ed.) (1989) *Counterurbanization*, London: Edward Arnold.

City of Birmingham (1986) *Land for Industry*, Birmingham: Development Department.

Clark, C. (1967) *Population Growth and Land Use*, London: Macmillan.

Clawson, M. (1972) *America's Land and Its Uses*, Baltimore: Johns Hopkins University Press.

Coleman, A. (1978) 'Planning and land use', *Chartered Surveyor*, 111: 158–63.

Deane, G. (1986) 'Statistics measure postwar change', *Town and Country Planning*, December: 346–7.

Department of the Environment (1978) *Developed Areas 1969: A Survey of England and Wales from Air Photography*, London: HMSO.

—— (1979) *Study of the Availability of Private House-building Land in Greater Manchester 1978–1981*, London: HMSO.

—— (1984) *First Report from the Environment Committee, Session 1983–8, Green Belt and Land for Housing*, vol. 2, Cmnd 8345, London: HMSO.

—— (1986) *Land Use Change in England Statistical Bulletin 86/1*, London: Government Statistical Service.

Department of the Environment (1987) *Land Use Change in England Statistical Bulletin 87/7*, London: Government Statistical Service.

—— (1988a) *Land Use Change in England Statistical Bulletin 88/5*, London: Government Statistical Service.

—— (1988b) *Urban Land Markets in the UK*, London: HMSO.

—— (1989) *Land Use Change in England Statistical Bulletin 89/5*, London: Government Statistical Service.

Dickenson, G.C. and Shaw, M.G. (1982) 'Land use in Leeds 1957–76: two decades of change in a British city, *Environmental and Planning*, A14: 343–58.

Dill, H.W. and Otte, R.C. (1971) *Urbanization of Land in the NE United States*, Economic Research Service Report 485, Washington DC: Government Printing Office.

Englestoft, S. (1989) 'Danish market towns – land use and development', *Scandinavian Housing and Planning Research*, 6: 1–16.

Fothergill, S. and Gudgin, G. (1982) *Unequal Growth: Urban and Regional Employment Change in the UK*, London: Heinemann.

Fothergill, S., Kitson, M. and Monk, S. (1985) *Urban Industrial Change*, Inner Cities Research Programme, London: HMSO.

Hauser, J.G. (1982) *Land Use Structure and Change in North America and the EEC*, Occasional Paper no. 6, Wye College, University of London.

Himiyama, Y. (1985) 'The use of Japanese land use maps at 1:25,000', *Land Use Policy*, 2: 278–88.

Himiyama, Y. and Jitsu, K. (1988) 'Recent achievements in land use studies, *Geographical Review of Japan*, B61: 99–110.

Home, R.K. (1985) 'Forecasting housing land requirements', *Land Development Studies* 2: 19–34.

Hooper, A., Pinch, P. and Rogers, S. (1988) 'Housing land availability', *Journal of Planning Law and Environmental Law*, April, 225–39.

Horner, A.A. (1987) 'Population and land use change in the Dublin region', in A.A. Horner and A.J. Parker (eds) *Geographical perspectives in the Dublin region*, Geographical Society of Ireland, Special Publication, no. 2, 70–95.

International Regional Science Review (1982) 7: 3, Special issue on regional development and the preservation of agricultural land.

Jackson, R.H. (1981) *Land use in America*, V.H. Winson/Edward Arnold.

Jones, A.R. (1974) 'An analysis of the major features of urban land use and land provision in cities and towns in England and Wales in about 1960', University of London, unpublished Ph.D. thesis.

Kivell, P.T. (1975) 'Postwar urban residential growth in North Staffordshire', in A.D.M. Phillips and B.J. Turton (eds) *Environment, Man and Economic Change*, London: Longman, 441–58.

Law, C.M. (ed.) (1988) *The Uncertain Future of the Urban Core*, London: Unwin Hyman.

Lecoin, J.-P. (1988) 'Paris and the Ile de France', in H. van der Cammen (ed.) *Four Metropolises in W. Europe*, Assen/Maastrier: Van Gorcum, 61–176.

Leven, C.L. (1978) 'Growth and non growth in metropolitan areas and the emergence of Polycentric Metropolitan form, *Papers, Regional Science Association*, 48: 101–12.

Lever, W. (ed.) (1987) *Industrial Change in the UK*, Harlow: Longman.

London Strategic Policy Unit (1987) *Land for Industry – the Need for Industrial Land in London until 1990*, London: Economic Regeneration Team.

Loveless, J. (1989) 'Commercial and industrial land', in R. Banks (ed.) *Costing the Earth*, London: Shepheard Walwyn, 86–102.

McDonald, J.F. (1985) 'The intensity of land use in urban employment sectors, Chicago 1960–1970', *Jnl. of Urban Economies*, 18: 261–77.

Muller, P. (1981) *Contemporary Suburban America*, Englewood Cliffs, NJ: Prentice-Hall.

OECD (1985) *Compendium of Environmental Data*, Paris.

Robson, B. (1988) *Those Inner Cities*, Oxford: Clarendon Press.

Short, J.R. (1989) 'Yuppies, yuffies and the new urban order', *Transactions of the Institute of British Geographers*, 14: 173–88.

Speake, J. (1991) 'East Manchester: a study of inner city industrial decline and regeneration', University of Keele, unpublished Ph.D. thesis.

Spencer, K. *et al.* (1986) *Crisis in the Industrial Heartland: a Study of the West Midlands*, Oxford: Clarendon Press.

United Nations Organisation (1976) *Global Review of Human Settlements*, Oxford: Pergamon Press.

—— (1987) *Global Report on Human Settlements*, Habitat: Oxford University Press.

United Nations (1989) *Annual Bulletin of Housing and Building Statistics 1980–88*, vol. 32, UNO: New York.

US Bureau of the Census (1988) *Statistical Abstract of the United States, 1988*, Washington DC: Government Printing Office.

US Soil Conservation Service (1971) *Basic Statistics of the National Inventory of Soil and Water Conservation needs*, Statistical Bulletin 461, Washington DC: Government Printing Office.

—— (1979) *National Summaries of the 1977 Resource Inventories*, Washington DC: Government Printing Office.

Volkman, N. (1987) 'Vanishing land in the USA', *Land Use Policy*, 4, 1: 14–30.

Volksblatt (1990) 18 May Seite 7.

Wilder, M. (1985) 'Site and situation determinants of land use change: an empirical example', *Economic Geography*, 4: 322–44.

Wood, J.S. (1988) 'Suburbanization of Center City, *Geographical Review*, 78, 3: 325–9.

Young, R.W. and Schoolmaster, F.A. (1985) 'Land use change around Dallas–Fort Worth Airport', *Papers and Proceedings of Applied Geography Conference*, (ed.) J.W. Frazier, SUNY, Binghampton, 286–97.

Zeimetz, K.D. *et al.*. (1976) *Dynamics of Land Use in Fast Growth Areas*, US Department of Agriculture Economic Research Service Report 325, Washington DC: Government Printing Office.

5 LAND OWNERSHIP

Adams, C.D. and May, H. (1990) 'Land ownership and land-use planning', *The Planner*, 76, 38: 11–14.

Adams, C.D., Baum, A.E and McGregor, B.D. (1988) 'The availability of land for inner city development: a case study of inner Manchester', *Urban Studies*, 25: 62–76.

Adrian, C. and Simpson, R. (1986) 'Asian investment in Australian capital city property markets', *Environment and Planning*, A, 18: 323–40.

Allinson, J. (1988) 'The community land scheme: a study in dichotomy and consensus', *The Planner* 74, 9: 29–32.

Ambrose, P. and Colenutt, R. (1975) *The Property Machine*, Harmondsworth: Penguin.

Atmer, T. (1987) 'Land banking in Stockholm', *Habitat International* 11, 1: 47–55.

Bailey, A.J. (1987) 'Public land disposals in the Metropolitan green belt', Bristol Polytechnic Department of Town and Country Planning, Working Paper 19.

Balchin, P.N. and Bull, G.H. (1987) *Regional and Urban Economics*, London: Harper & Row.

Barrett S.M. and Healey, P. (1985) *Land Policy: Problems and Alternatives*, Aldershot: Gower.

Barrett, S.M., Stewart, M. and Underwood, J. (1978) 'The land market and development process: a review of research and policy', 2, Occasional Paper School of

Advanced Urban Studies, University of Bristol.

Blitz, E., Houdijk, T. and Maas, A. (1988) 'Real estate development in The Hague', *Prospect* 2, 16–18.

Bracewell-Milnes, B. (1982) *Land and Heritage: the Public Interest in Personal Ownership*, Hobart Paper 13, London: Institute of Economic Affairs.

Bryant, R.W.G. (1972) *Land: Private Property, Public Control*, Montreal: Harvest House.

—— (1976) 'Problems of implementation: an overview', in D. Kehoe, *et al.* (eds) *Public Land-ownership: Frameworks for Evaluation*, 68–81, Massachusetts: Lexington Books.

Carr, J. and Smith, L. (1975) 'Public land banking and the price of land', *Land Economics* 51, 4: 316–30.

Castells, M. (1977) *The Urban Question*, London: Methuen.

Clawson, M. (1971) *Suburban Land Conversion in the United States*, Baltimore: Johns Hopkins University Press.

Conzen, M. (1960) *Alnwick: A Study of Town Plan Analysis*, Institute of British Geographers, Monograph no. 27.

Cox, A. (1984) *Adversary Politics and Land*, Cambridge: Cambridge University Press.

Cummings, S., Koebel, C.T. and Whitt, J.A. (1988) 'Public–private partnerships and public enterprise', *Urban Resources* 5, 1: 35–6.

Daly, M.T. (1984) 'The revolution in international capital markets: urban growth and Australian cities, *Environment and Planning* A, 16: 1003–20.

Dear, M. and Scott, A.J. (eds) (1981) *Urbanization and Urban Planning in Capitalist Society*, London: Methuen.

Denman, D.R. (1974) 'Land nationalization – a way out?', *Government and the Land, IEA Readings 13*, London: Institute of Economic Affairs.

—— (1978) *The Place of Property*, Berkhamstead: Geographical Publications.

Dowrick, F.E. (1974) 'Public ownership of land – taking stock 1972–73', *Public Law*, 10–24.

Duncan, S. (1989) 'Development gains and housing provision in Britain and Sweden', *Transactions Institute of British Geographers*, 14, 2: 157–72.

Dyos, H. (1968) *The Study of Urban History*, London: Edward Arnold.

Edwards, M. and Lovatt, D. (1980) *The Inner City in Context, vol. 1. Understanding Urban Land Values: A Review*, Report for Social Science Research Council, London.

Evans, A. (1988) *No Room! No Room! The Costs of the British Town and Country Planning System*, Occasional Paper 79, Institute of Economic Affairs, London.

Flatt, A. (1982) 'Achieving effective systems of land cadastres, evaluation and title registration', in M. Cullen, and S. Wolley, (eds) *World Congress on Land Policy*, Massachusetts: Lexington Books.

Fleming, J.S. and Little, I.M.D. (1975) *Why We Need a Wealth Tax*, London: Methuen Pamphlets.

Goodchild, R. and Munton, R. (1985) *Development and the Landowner*, London: Allen & Unwin.

Hall, P. (1976) 'A review of policy alternatives', in D. Kehoe, *et al.* (eds) *Public Land-ownership: Frame-works for Evaluation*, 46–56, Massachusetts: Lexington Books.

Hallett, G. (1977) *Housing and Land Prices in Britain and Germany*, London: Macmillan.

—— (1979) *Urban Land Economics*, London: Macmillan.

Hamilton, S.W. and Baxter, D.E. (1977) 'Government ownership and the price of land', in L.B. Smith and M. Walker (eds) *Public Property? The Habitat Debate Continued*, Fraser Institute, Vancouver BC: 75–118.

Hansard (1988) Written Answers, 12 Dec. vol. 143, cols 405–7.

Harrison, A., Tranter, R.B. and Gibbs, R.S. (1977) *Landownership by Public and Semi-public Institutions in the United Kingdom*, University of Reading: Centre for Agricultural Strategy.

Kehoe, D., Morley, D., Proudfoot, S.B. and Roberts, N.A. (eds) (1976) *Public Land Ownership: Frame-works for Evaluation*, Massachusetts: Lexington Books.

Kilmartin, L. and Thorns, D.C. (1978) *Cities Unlimited: the Sociology of Urban Development in Australia and New Zealand*, Sydney: Allen & Unwin.

Kivell, P.T. and McKay, I. (1988) 'Public ownership of urban land', *Transactions of Institute of British Geographers*, 13: 165–78.

Knobel, C. (1988) 'The men who bought London', *Management Today*, August, 62–6.

Lefebvre, H. (1977) 'Reflections on the politics of space', in R. Peet (ed.) *Radical Geography*, Chicago: Maaroufa Press.

Lloyd, M.G. (1989) 'Land development and the free market lobby', *Scottish Planning Law and Practice* 26: 8–10.

Lock, D. (1989) 'New communities in Britain: the role of the private sector', *The Planner* 75, 2; 33–6.

Markusen, J.R. and Scheffman, D.T. (1977) 'Ownership concentration in the urban land market: analytical foundations and empirical evidence', in L.B. Smith and M. Walker (eds) *Public Property, The Habitat Debate Continued*, Vancouver: Fraser Institute, 147–76.

Massey, D. and Catalano, A. (1978) *Capital and Land: Landownership by Capital in Great Britain*, London: Edward Arnold.

Meyerson, M. and Banfield, E.C. (1964) *Politics, Planning and the Public Interest*, London: Collier Macmillan.

Montgomery, J.R. (1987) 'The significance of public land ownership', *Land Use Policy* 4, 1: 42–50.

Mortimore, M.J. (1969) 'Land ownership and urban growth in Bradford', *Transactions Institute of British Geographers*, 46: 105–19.

Needham, B. (1983) 'Local government policy for industrial land in England and the Netherlands', in S. Barrett and P. Healey, (eds) *Land Policy: Problems and Alternatives*, 298–307, Aldershot: Gower.

Neutze, M. (1989) 'A tale of two cities', *Scandinavian Housing and Planning Research*, 6, 4: 189–99.

Norton-Taylor, R. (1982) *Whose Land Is It Anyway?* London: Turnstone.

Nowlan, D. (1977) 'The land market: how it works, in L.B. Smith and M. Walker (eds) *Public Property, the Habitat Debate Continued*, 147–76, Vancouver: Fraser Institute.

Ratcliffe, J. (1976) *Land Policy*, London: Hutchinson.

Roberts, N.A. (1977) *The Government Land Developers*, Massachusetts: Lexington Books.

Sant, M. (1980) 'Acquisition, management and disposal of land', *Town and Country Planning*, 49: 146.

Saunders, P. (1981) *Social Theory and the Urban Question*, London: Hutchinson.

Self, P. (1988) 'A new international city for Australia', *Town and Country Planning* 57, 11: 310–11.

Shoup, D.C. (1983) 'Intervention through property taxation and public ownership', in H.B. Dunkerley (ed.) *Urban Land Policy*, 132–70, New York: Oxford University Press.

Simmie, J.M. (1978) *Citizens in Conflict, the Sociology of Town Planning*, London: Hutchinson.

Slaughter, J.C. (1973) 'Is it time for a change in land tenure systems?', *The Developer*, October 179–80.

Sutherland, D. (1988) *The Landowners*, London: Muller.

Thrift, N. (1986) 'The internationalisation of producer services and the integration of the Pacific Basin property market', in N. Thrift and M. Taylor (eds) *Multinationals and the Restructuring of the World Economy*, London: Croom Helm.

Ward, D. (1962) 'The pre-urban cadaster and the urban pattern of Leeds', *Ann. Ass. Am. Geogr*, 52: 150–66.

Wharf, B. (1988) 'Japanese investment in the New York metropolitan region', *Geographical Review*, 78, 3: 257–71.
White, P. (1986) 'Land availability, land banking and the price of land for housing: a review of recent debates, *Land Development Studies* 3: 101–11.

6 LAND POLICY

Ambrose, P. (1986) *Whatever Happened to Planning?* London: Methuen.
Ambrose, P. and Colenutt, R. (1975) *The Property Machine*, Penguin.
Balchin, P.N. and Bull, G.H. (1987) *Regional and Urban Economics*, London: Harper & Row.
Barrett, S. and Healey, P. (eds) (1985) *Land Policy Problems and Alternatives*, Aldershot: Gower.
Barrett, S., Stewart, M. and Underwood, J. (1978) 'The land market and the development process', Occasional Paper no. 2, School for Advanced Urban Studies, Bristol University.
Bosselman, F. and Callies, D. (1972) *The Quiet Revolution in Land Use Control*, Washington: Government Printing Office.
Bourne, L.S. (1967) 'Private redevelopment of the central city', University of Chicago, Geography Department, Research Paper 112.
Boyer, M.C. (1981) 'National land use policy: instrument and product of the economic cycle', in J.I. de Neufville (ed.) *The Land Use Policy Debate in the United States*, New York: Plenum Press, 109–126.
Calavita, N. (1984) 'Urbanization, public control of land use and private ownership of land: the development of Italian Planning Law', *Urban Lawyer*, 16, 3: 459–88.
Cherry, G. (1991) 'Today's issues in a twentieth century perspective', *The Planner*, 77, 1: 7–10.
Chisholm, M. and Kivell, P.T. (1987) *Inner City Wasteland*, London: Institute of Economic Affairs.
Clarke, W.A. (1974) 'The impact of property taxation on urban spatial development', Report no. 187, Los Angeles: Institute of Government and Public Affairs, University of California.
Cox, A. (1984) *Adversary Politics and Land*, Cambridge: Cambridge University Press.
Darin-Drabkin, H. (1977) *Land Policy and Urban Growth*, Oxford: Pergamon Press.
Delafons, J. (1991) 'Planning in the USA – a better system?' *The Planner*, 77, 4: 7–8.
de Neufville, J.I. (1981) *The Land Use Policy Debate in the USA*, New York: Plenum Press.
Department of the Environment (1988) *Urban Land Markets in the United Kingdom*, London: HMSO.
Dowall, D.E. (1989) 'Land use policy in the USA', *Land Use Policy*, 6, 1: 11–30.
Drewett, J.R. (1973) 'The developer's decision process', in P. Hall, *et al.* (eds) *The Containment of Urban England*, vol. II, 163–245, London: Allen & Unwin.
Dunkerley, H.B. (ed.) (1983) *Urban Land Policy*, Oxford: University Press.
Dunlap, R. (1987) 'Polls, pollution and politics revisited: public opinion on the environment in the Reagan era', *Environment*, 29, 4: 6–11, 32–7.
Frieden, B.J. (1990) 'Center city transformed', *Journal of American Planning Association*, 50, 4: 423–8.
George, H. (1879) *Progress and Poverty. Centenary Edition*, 1979, New York: Robert Schlackenback Foundation.
Goodchild, R. and Munton, R. (1985) *Development and the landowner*, London: Allen & Unwin.
Haigh, N. (1989) 'The European Community and land use – an incoming tide', *Journal of Planning and Environment Law*, Occasional Paper 16, 58–64.

Hall, P. (1974) 'The containment of urban England', *Geographical Journal*, 140, 3: 386–408.

Hall, P., Gracey, H., Drewett, R. and Thomas, R. (1973) *The Containment of Urban England*, vols I and II, London: Allen & Unwin.

Hallett, G. (1977) *Housing and Land Policies in W. Germany and Britain*, London: Macmillan.

—— (ed.) (1988) *Land and Housing Policies in Europe and the USA*, London: Routledge.

Harrison, F. (1983) *The Power in the Land*, London: Shepheard-Walwyn.

Healey, P., McNamara, P., Nelson, M. and Doak, A. (1988) *Land Use Planning and the Mediation of Urban Change*, Cambridge: Cambridge University Press.

Holland, D.M. (1971) *The Assessment of Land Value*, Madison: University of Wisconsin Press.

Keating, W.D. and Krumholz, N. (1991) 'Downtown plans of the 1980s: the case for more equity in the 1990s', *Journal of American Planning Association*, 57, 2: 136–52.

Keogh, G. (1985) 'The economics of planning gain', in S. Barrett and P. Healey (eds), *Land Policy: Problems and Alternatives*, Aldershot: Gower, 203–28.

Knapp, G.J. (1987) 'Self interest and voter support for Oregon's land use controls', *Journal of the American Planning Association*, 53, 1: 92–7.

Lee, D.B. (1981) 'Land use planning as a response to market failure', in J.I. de Neufville, *The Land Use Policy Debate in the United States*, New York: Plenum Press, 149–66.

Lichfield, N. (1956) *Economics and planned development*, London Estates Gazette.

Lichfield, N. and Darin-Drabkin, H. (1980) *Land Policy in Planning*, London: Allen & Unwin.

Markusen, A. (1981) 'Introduction to the political economy perspective', in J.I. de Neufville (ed.) *The Land Use Policy Debate in the United States*, 103–8.

Massey, D. and Catalano, A. (1978) *Capital and Land*, London: Edward Arnold.

Meck, S. (1990) 'From high minded reformism to hard boiled pragmatism: American city planning faces the next century', *The Planner*, 76, 6: 11–14.

Miliband, R. (1973) *The State in Capitalist Society*, London: Quartet.

Popper, F.J. (1988) 'Understanding American land use regulation since 1970: a revisionist interpretation, *Journal of the American Planning Association*, 54, 3: 291–301.

Punter, J.V. (1989) 'France', in H.W.E. Davies, D. Edwards, A.J. Hooper and J.V. Punter (eds) *Planning Control in Western Europe*, Department of the Environment, London: HMSO.

Ratcliffe, J. (1976) *Land Policy: An Exploration of the Nature of Land in Society*, London: Hutchinson.

Richman, R.L. (1965) 'The theory and practice of site valuation in Pittsburgh', Proceedings of the 57th Annual Conference on Taxation', Harrisburg, Pennsylvania: National Tax Association.

Royal Town Planning Institute, (1985) *Land Policy in Britain*, discussion document prepared by the Land Policy Working Party, London: RTPI.

Rydin, Y. (1986) *Housing Land Policy*, Aldershot: Gower.

Wakeford, R. (1990) *American Development Control*, London: HMSO.

Woodruff, A.M. and Ecker-Racz, L.L. (1969) 'Property taxes and land use patterns in Australia and New Zealand', in A. Becker (ed.), *Land and Building Taxes*, Madison: University of Wisconsin Press.

7 VACANT AND DERELICT LAND

Adams, C.D., Baum, A.E. and McGregor, B.D. (1985) 'The influences of valuation practices upon the price of inner city land', *Land Development Studies*, 2, 157–73.

—— (1987) 'Vacant urban land – the planner's responsibility?', *The Planner*, June, 31–4.

—— (1988) 'The availability of land for inner city development: a case study of inner Manchester, *Urban Studies*, 25: 62–76.

Aitken, (1988) 'Land renewal in South Limburg', *The Planner*, 72, 9: 25–7.

Anderson, J., Duncan, S. and Hudson, R. (1983) *Redundant Spaces in Cities and Regions*, IBG Special Publication no. 15, London: Academic Press.

Bacon, E. (1976) 'Total national commitment needed to restore abandoned property', *Urban Land*, 35: 3–4.

Ball, R.M. (1989) 'Vacant industrial premises and local development, a survey, analysis and policy assessment of the problem in Stoke on Trent', *Land Development Studies*, 6: 105–28.

Baum, A.E. (1985) 'Land availability for inner city redevelopment', Department of Land Management, University of Reading, vol. II.

Bruton, M.J. and Gore, A. (1981) 'Vacant urban land in South Wales, *Local Government Review*, 14/3/81.

Burrows, J. (1978) 'Vacant urban land – a continuing crisis', *The Planner*, January, 7–9

Cameron, G., Monk, S. and Pearce, B.J. (1988) *Vacant Urban Land, A Literature Review*, London: Department of the Environment.

Carlsson, C.J. and Duffy, R.J. (1985) 'Cincinnati takes stock of its vacant land', *Planning* 51, 11: 22–6.

Chaix, R. (1989) 'Friches industrielles et réaffectations en Ile de France, Evolution 1985–88', *Hommes et Terres du Nord*, 4: 320–4.

Chaline, C. (1988) 'La reconversion des espaces fluvio–portuaires dans les grandes métropoles', *Annales de Geographie*, 544: 695–715.

Chisholm, M. and Kivell, P.T. (1987) *Inner City Wasteland*, London: Institute of Economic Affairs.

Civic Trust (1988) *Urban Wasteland Now*, Civic Trust: London.

Clout, H. (1988) 'French approach for coping with dereliction', *Town and Country Planning*, 57: 1.

Coleman, A. (1980) 'The death of the inner city, cause and cure', *The London Journal* 6: 1.

—— (1982) 'Dead space in the dying inner city', *International Journal of Environmental Studies*, 19: 103–7.

Couch, C. (1989) 'Vacant and derelict land in France', *Land Development Studies*, 6: 183–99.

Couch, C. and Herson, J. (1986) 'The Grandstuckfond–Ruhr: a system for managing derelict land', *The Planner* 72, 9: 23–4.

Dallas, (no date) 'Infilling potential in the Dallas inner city', Draft Special Report, Department of Housing and Urban Rehabilitation, Dallas, Texas.

Dawson, A.H. (1979) 'Empty land in the Scottish city', *Sonderdruek aus der Zeitschrift, Wirtschaftz geographische studien*, Heft, 5, 3: 46–67.

Department of the Environment (1987) *Evaluation of Derelict Land Grant Schemes*, Inner Cities Research Programme, London: HMSO.

—— (1988) *Land Use Change in England*, Statistical Bulletin 88/5, London: Government Statistical Service.

—— (1989) *A Review of Derelict Land Policy*, London: HMSO.

—— (1991) *Survey of Derelict Land in England*, 1988, London: HMSO.

REFERENCES

Ernecq, J.M. (1988) 'Quelques aspects de la politique de l'environment dans une région de tradition industrielle', in F. Joyce and G. Schneider (eds) *Development in the Regions of the European Community*, Aldershot: Avebury.

Falk, N. (1985) 'Our industrial heritage, a resource for the future', *The Planner* 71, 10: 13–16.

Fothergill, S., Kitson, M. and Monk, S. (1983) *Industrial Land Availability in Cities, Towns and Rural Areas*, Department of Land Economy, University of Cambridge.

Fox, T. (1989) 'Using vacant land to reshape American cities', *Places* 6, 1: 78–81.

HM Government (1984) *Green Belt and Land for Housing*, First Report of the House of Commons Environment Committee, vol. 1, Para 78, HMSO: London.

Harvey, D. (1974) 'Class-monopoly rent, finance, capital and the urban revolution', *Regional Studies*, 8, 3/4: 239–55.

Holzer, T.L. (1989) 'State and local response to damaging land subsidence in US urban areas', *Engineering Geology*, 27: 449–66.

Hoyle, B.S., Pinder, D.A. and Husain, M.S. (eds) (1988) *Revitalizing the waterfront*, London: Bellhaven.

Kivell, P.T. (1987) 'Derelict land in England, policy responses to a continuing problem', *Regional Studies* 21, 3: 265–73.

—— (1989) 'Vacant urban land: intervention or the market?', *The Planner*, August: 8–9.

Lefebvre, H. (1970) *La Révolution Urbaine*, Paris: Gallimard.

Loveless, J. (1987) *Why Wasteland? Towards an Urban Renaissance*, Adam Smith Institute Research Ltd: London.

Malezieux, J. (1987) 'Réanimation de friches industrielles en banlieue Parisienne, Milieux, Villes et Régions, Actes du 112e Congres National des Societies Savantes', Lyons, 174–94.

Martin, R. and Rowthorn, R. (1986) *The Geography of Deindustrialisation*, Cambridge: Cambridge University Press.

Massey, D. and Meeghan, R. (1982) *The Anatomy of Job Loss*, London: Methuen.

Moor, N. (1985) 'Inner city areas and the private sector', in S.M. Barrett and P. Healey (eds) *Land Policy: Problems and Alternatives*, Aldershot, Gower.

Nicholson, D.J. (1984) 'Public ownership of vacant urban land', *The Planner* 70, 1: 18–20.

Northam, R.M. (1971) 'Vacant urban land in the American City', *Land Economics* 47: 345–55.

Partington, M. (1986) *Wasted Land, a Phase in the Life Cycle of British Cities*, London: Polytechnic of Central London.

Pinder, D. and Rosing, K.E. (1988) 'Public policy and planning of the Rotterdam Waterfront', in B.S. Hoyle *et al. Revitalizing the waterfront*, 114–45, London: Bellhaven.

Portland (1978) *Vacant Land Report*, Bureau of Planning, Portland, Oregon.

Rose, E.A. (ed.) (1986) *New Roles for Old Cities*, Aldershot: Gower.

Sinn, H.W. (1986) 'Vacant land and the role of government intervention', *Regional Science and Urban Economics* 16, 3: 353–85.

Speake, J. (1991) 'East Manchester: a study of inner city industrial decline and regeneration', unpublished Ph.D. thesis, University of Keele.

Stevens, B. (1988) 'New lease of life for old docklands', *Dock and Harbour Authority*, 69, 801: 1–3.

Thomas, F. and Cretin, C. (1987) 'La Réhabilitation des friches industrielles de l'agglomération Stephanoise Milreux, Villes et Régions, Actes du 112e Congrès National des Sociétés Savantes, Lyons: 257–66.

Thorpe and Partners (1982) *Industrial Floorspace*, London.

Watson, G. (1987) 'Recycling disused industrial land in the Black Country', Department of Town Planning Working Paper no. 97, Oxford Polytechnic.

Watts, H.D., Smithson, P.A. and White, P.E. (1989) *Sheffield Today*, Sheffield: Department of Geography, University of Sheffield.

8 SOME CONCLUSIONS

Ash, M. (1969) *Regions of Tomorrow, Towards the Open City*, London: Evelyn, Adams and Mackay.

Audirac, I., Shermyen, A.H. and Smith, M.T. (1990) 'Ideal urban form and visions of the good city', *Journal of the American Planning Association*, 56, 4: 470–82.

Bell, D. (1974) *The Coming of Post Industrial Society: A Venture in Social Forecasting*, London: Heinemann.

Berry, B.J.D. and Cohen, Y.S. (1973) 'Decentralization of commerce and industry', in L.H. Masotti and J.K. Hadden (eds) *The Urbanization of the Suburbs*, Urban Affairs Annual Review, 17, Beverley Hills: Sage.

Boyce, R.R. (1966) 'The edge of the metropolis: the wave analog approach', *The Geographer and the Environment*, British Columbia Geographical Series 7, Department of Geography, University of British Columbia; 31–40.

Breheny, M. (1991) 'Contradictions of the Compact City', *Town and Country Planning*, 60, 1: 21.

Bryant, C.R., Russwurm, L.H. and McLellan, A.G. (1982) *The City's Countryside*, London: Longman.

Cadman, D. and Payne, G. (eds) (1989) *The Living City: Towards a Sustainable Future*, London: Routledge.

Castells, M. (1990) *The Informational City: Information Technology, Economic Restructuring and the Urban Regional Process*, Oxford: Basil Blackwell.

Champion, A.G. (ed.) (1989) *Counterurbanization: the changing pace and nature of population deconcentration*, London: Edward Arnold.

Champion, A.G. and Townsend, A.R. (1990) *Contemporary Britain*, London: Edward Arnold.

Cherry, G. (1991) 'Today's issues in a twentieth century perspective', *The Planner*, 1: 7–10.

Cohen, S. and Zysman, J. (1987) *Manufacturing Matters: The Myth of the Post Industrial Economy*, New York: Basic Books.

Commission of the European Communities (1990) *Green Paper on the Urban Environment*, COM (90) 218, Brussels.

Cooke, P. (ed.) (1989) *Localities: The Changing Face of Urban Britain*, London: Unwin Hyman.

Drucker, P. (1971) *The Age of Discontinuity: Guidelines to Our Changing Society*, London: Pan.

Elkin, T., McLaren, D. and Hillman, M. (1991) *Reviving the City – towards Sustainable Urban Development*, London: Friends of the Earth and Policy Studies Institute.

Fothergill, S., Gudgin, G., Kitson, M. and Monk, A. (1986) 'The deindustrialisation of the city', in Martin, R. and Rowthorn, B. (eds) *The Geography of Deindustrialization*, London: Methuen.

Fulton, W. (1986) 'Offices in the dell', *Planning*, 57, 2: 13–17.

Gappert, G. (1985) 'Urban issues in an advanced industrial society', in Hall, P. and Markusen, A. (eds) *Silicon Landscapes*, Boston: Allen & Unwin; 424–34.

Garreau, J. (1988) 'Edge Cities', *Landscape Architecture*, 78, 8: 48–55.

Gershuny, J. and Miles, I. (1983) *The New Service Economy: the Transformation of Employment in Industrial Societies*, London: Frances Pinter.

Goodchild, B. (1990) 'Planning and the modern/postmodern debate', *Town Planning Review*, 61, 2: 119–37.

Gottdeiner, M. (ed.) (1986) *Cities in Stress: A New Look at the Urban Crisis*, Urban Affairs Annual Review, 30, Beverley Hills: Sage.

Gottman, J. (1983) *The Coming of the Transactional Society*, Baltimore: University of Maryland.

Hall, P. (1985) 'A problem with its roots in the distant past', *Town and Country Planning*, 54, 2: 40–3.

——— (1987) 'The geography of the post industrial economy', in Brotchie, J.F., Hall, P. and Newton, P.W. (eds.) *The Spatial Impact of Technological Change*, London: Croom Helm: 3–17.

——— (1990) 'Transitional stress and the crisis of those without hope', *Town and Country Planning*, 59, 12: 331–3.

——— (1991) 'Three systems, three separate paths', *Journal American Planning Association*, 57, 1: 16–20.

Hall, P. and Markusen, A. (1985) *Silicon Landscapes*, Boston: Allen & Unwin.

Hardy, D. (1990) 'Farewell modern city', *Town and Country Planning*, 59, 12: 324–5.

Harris, B. (1987) 'Cities and regions in the electronic age', in Brotchie, J.F., Hall, P. and Newton, P.W., *The Spatial Impact of Technological Change*, London: Croom Helm; 394–405.

Hart, J.F. (1991) 'The perimetropolitan bow wave', *Geographical Review*, 81, 1: 35–51.

Harvey, D. (1985) *The Urbanization of Capital*, Oxford: Basil Blackwell.

——— (1989) *The Condition of Postmodernity*, Oxford: Blackwell.

Hay, D. (1990) 'On the development of cities', in Cadman, D. and Payne, G. (eds) *The Living City*, London: Routledge.

Henley Centre (1990) *Local Futures*, London: Henley Forecasting Centre.

Herington, J. (1984) *The Outer City*. London: John Wiley.

——— (1991) 'How to deal with urban shorelines', *Town and Country Planning*, 60, 4: 124–6.

Holloway, S.R. and Wheeler, J.O. (1991) 'Corporate headquarters relocation and changes in metropolitan corporate dominance 1980–7', *Economic Geography*, 67, 1: 54–74.

Jacobs, J. (1961) *The Death and Life of Great American Cities*, New York: Random House.

Johnson, J.H. (1974) *Suburban Growth*, London: John Wiley.

Krier, L. (1987) 'Tradition, modernity, modernism – some necessary explanations', *Architecture Design Profile*, 65.

Lyotard, J.-F. (1984) *The Post-Modern Condition*, Manchester: Manchester University Press.

May, G. (1990) 'Information technology and the future of our cites', *Town and Country Planning*, 59, 1: 20–1.

Newman, P. (1990) 'The Search for the Good City', *Town and Country Planning*, 59, 10: 272–4.

Pacione, M. (1990) 'Development Pressures in the Metropolitan Fringe', *Land Development Studies*, 7: 69–82.

Pivo, G. (1990) 'The net of mixed beads: suburban office development in six metropolitan regions', *Journal American Planning Association*, 56, 4: 457–69.

Steeley, G. (1990) 'An Aegean of cities', *Town and Country Planning*, 59, 12: 333.

Touraine, A. (1974) *The Post Industrial Society*, London: Wildwood House.

Ward, C. (1990) 'Old prophets of new city regions', *Town and Country Planning*, 59, 12: 329–30.

INDEX